Quantum Prayer: It's Already Answered!
Quantum Reality: It Already Exists!

To Glenda & Paul,

With Much Love & Gratitude!

Enjoy!!!

Always,

Cathy

Quantum Prayer: It's Already Answered!
Quantum Reality: It Already Exists!

Cathy Combs

Quantum Prayer: It's Already Answered!

For updated contact information: www.cathycombs.com

ISBN 978-0-9843800-0-8

Printed in the United States Of America

Dedicated to

Reclaiming The Awareness Of

The Unitive Principle In All Of Life

Acknowledgments

As always in a project of this magnitude there are many people to acknowledge and thank. My first acknowledgment is to my parents. Their singular devotion to education and service colors my world to this day. I would not be who I am without their influence and without their example. Thank you everlastingly mom and dad. I love you both.

Next, I want to thank the many people who read my manuscript in toto or in part. Just by your willingness to be involved at this level I was greatly encouraged and buoyed all the way through this 2-1/2 year process. Thank you Dawn O'Malley for being the first one to offer unabashed support of what I was saying, and ironically the one who brings this project full circle offering the technical expertise to make this a finished product.

Thank you to Rev. Nancy Jerome Lytle for reading my manuscript while you were still working full time wearing many hats.

A special thank you to Rev. Bob Barth who I asked to read my manuscript as an editor and comment on the clarity and content. Your help, support, and humor were invaluable.

A special thank you to Terri Evans who I also asked to read my manuscript as an editor for clarity, readability, and all the other details: grammar, punctuation, typos. Your glowingly emphatic thumbs up made me laugh.

I called my readers my Angels Embodied Team. That is certainly what you felt like in this process. Thank you all: Bukeka Newby Shoals, Teri Wilder, Keith Mallory, Jan Stromseth. I feel it is important to acknowledge that every one of my readers read my manuscript while they were working full time wearing many, many other hats. I appreciate your gracious support. I hope I haven't forgotten to mention any of my readers. I wish to thank you as well.

A special thank you to Deb Mathine who so cheerfully and graciously served as my layout and design artist. Beautiful job. Thank you.

I want to acknowledge my many, many friends across the country who have said, "Let me know the minute it's done. I'll market it here," and my many minister friends who have already invited me to speak in their ministries. I greatly appreciate your love and generosity. Thank you so much!!!!!!!!!!

Lastly, I want to especially acknowledge two people who have been lifelong Angels Embodied to me personally: my sister, Vicky, and my dear friend, Sharon Drenner. Thank you both for your very heartfelt love and support.

Table Of Contents

Quantum Prayer: It's Already Answered!
Quantum Reality: It Already Exists!
by Cathy Combs

Let it clearly be said: For this I have come: To enlarge our understanding of prayer and thus enlarge our definition of *God.* This is the discussion that rocks my world! This is the transformative discussion that is the very essence of Joy in my life. I am writing this book as a profound thank you to all the progressive enlightened spiritualities that understand that *God* is so very much more than the "angry, judgmental, capricious bearded white man in the sky image" so many of us have grown up with for the past 5,000 years. I am so very grateful for my intuitive childhood understanding that I am innately good; that I didn't have to go to a building to find *God*. I am so very grateful for my adult understanding that *God* **is everything**!!! That is the focus of this book.

My understanding of what prayer is comes out of my understanding of what *God* is. My current understanding is radically different than anything I was taught as a child. My current understanding of *God* and prayer is what continues to fuel my passionate devotion to making a profound difference in life. I am here to heal the divisiveness that rages in the world today. I am here to heal the rage, depression, and hopelessness I hear and see day in and day out. I live by example. I teach by example. I do everything in my power to walk my talk.

I am so profoundly grateful for the teachings of Unity that I found as a young adult when I first moved to California to "find my dream." The profound irony is I grew up five miles from the world headquarters of Unity. I was even on the grounds of Unity Village listening to a lecture by Dr. Viktor Frankl on

November 21 and 22, 1968 when I was a student at KU, and yet I never even heard of Unity until 1974 when my roommate came home exploding in joy shrieking, "Cath, Cath, Cath, you've just gotta go hear this woman!" My roommate had just heard "this woman" speaking at John F. Kennedy University, where I had just completed my graduate school training in 1973. I hadn't been inside a church in ten years at that point and I was thinking, "You've got to be kidding", but I went anyway! "This woman" turned out to be the Rev. Dr. Carol Ruth Knox, a Unity minister. My life changed that day!!! I cried and cried for weeks as I listened to that woman say what I had intuitively understood all my life: We are innately good!!!! We are **not** sinners!!!! We are the beloved of *God*, the Infinite Love of the Universe!!!!!! I couldn't go near that building without bursting into tears. I was so overcome with Joy. I had found my spiritual home where I felt loved for who I am.

Since that day in October of 1974 at Unity of Walnut Creek I have since found the teachings of Goddess spirituality, Gnosticism, EFT, the Abraham teachings, HeartMath, Nonviolent Communication, and the spiritual scientific revelations of quantum physics, spiral dynamics, holographic universes, parallel universes, and beyond, all of which have added so much depth, happiness, joy, confidence and understanding to my life. I indeed found my dream and continue to live it out to this very day! For these reasons and so many more I am passionately devoted to "being the change I wish to see in my world" as Gandhi invites us and exhorts us to be, because he also so passionately said, "An eye for an eye makes the whole world blind." I so passionately believe that with all my heart. My book is one effort to go in the direction of providing a vibrantly Joy-filled, enlightened option.

My book, "Quantum Prayer: It's Already Answered!", is my gift to a world I love so very much. Quantum Prayer is an extension of my understanding of how very much we are loved by a beneficent Universe that many call *God*. My book is an extension of my approach to life as the quintessential optimist knowing that *God* **is** everything! *God* is the Invisible Essence that underlies everything! That isn't what most of us have been taught but I have come to know it is true! I now understand that is why my whole life has been dedicated to personal empowerment for everyone. My soul knew this Truth long before I consciously knew it. Now that I consciously know that *God* is everything I'm living this Truth to the very best of my growing ability. I know this is the shared sacred task and gift for every one of us on this Earth.

I treasure the holistic viewpoint that echoes my understanding that *God* is everything. The holistic point of view holds that the Creator cannot be separate from Its creation! I believe with all my heart that the holistic point of view is the defining moment in the Universe. As we understand who we are from a holistic point of view we understand we **always** have a choice as to what we demonstrate in our lives. We are **never** passive victims for anything that happens. We are cocreators of every moment!

My major premise is we're all healers. Given the state of our world today, if we're only looking at appearances, that might seem to be an absurd statement. It isn't! I believe we're here to rise to the level of Consciousness that all the avatars of all the religions have risen to in their Earthly physical expressions. We're here to fully actualize the Truth that we are Love Incarnate. To say otherwise doesn't make sense to me. I have intuitively known these holistic principles since childhood. We are all innately good. *God* is within us and all around us. We don't have to go to a building to find *God*! *God* is everywhere present, every moment! *God* is Love Incarnate. So are we! We are here to fully actualize our true divine nature in physical reality! I am here to be the healer I am! I'm here to help others find their way to the healer within and to actualize it in this world, this lifetime!!! I'm here to join the vast array of individuals and perspectives detailing how to live our Oneness **as** the Divine here and now!!!

Lastly, even though I've divided my book into chapters you'll notice that many, if not all, of the ideas I'm sharing are in each chapter because they're so interrelated. These ideas are not distinct, separate units. They are all one! They flow inseparably **from** each other!

Chapter 1
What Is God?

What a surprise indeed when the answer to the question of all ages comes with such simple clarity that my understanding is, "Of course!" That's what happened today when the answer came through simple as can be: *God* is **everything**!!! I have long since understood the panentheistic view that *God* is **in** everyone, everything, every moment. That's old news to me. The new news came through with stunning clarity and simplicity. *God* is **in** everything because *God* **is** everything!" The ongoing surprise occurred for me today with even greater clarity **why** there is no separate Creator and then creation. Creation is *God* Itself made visible. There is no separate Creator orchestrating things from the outside! *God* **is** the process Itself!

This pantheistic understanding that *God* is everything that just dawned on me is an outgrowth of my longheld belief from my field of Psychology that the Creator cannot be separate from Its creation. This understanding makes it even more fascinating when considering whether there is any such thing as objective reality, or is it purely subjective as the Eastern mystics have said for so long. It is important in this discussion to clarify that objective reality is different than physical reality. There most certainly is a physical reality. Whether it is objective or purely subjective is another question entirely. Quantum physics, spiral dynamics, holographic universes, parallel universes, the Abraham teachings, Goddess spirituality are all adding other dimensions to this interdisciplinary question spanning spirituality, religion, science, psychology, holistic healing, and physics.

The book, "Natural Grace", co-written by noted British biologist Dr. Rupert Sheldrake and Rev. Dr. Matthew Fox, the voice of Creation Spirituality, is a wonderful expose on how science and spirituality intersect. I see clearly that the Cosmic Void that I call the Unseen Realm, the formless Substance from which all of creation comes into being, is *God* Itself. The Cosmic Void is Creativity Itself, the Creative Principle in Eternal Action.

We are Light. We are Energy. I now understand my belief that *God* is the only Presence and Power in the Universe, *God* the Good Omnipotent. *God* is **everything**! The question that immediately came after that is: "*God* is war? How can that be?" I didn't have an answer for a moment because that seemed incongruous to my understanding of *God* as Love Incarnate. Then I remembered the line from the Gnostic Gospel of Thomas, where Jesus is quoted as saying, "I came to bring you war, not peace."

I hear Jesus saying, "I came to shake you out of your complacency, your unawareness. Don't settle for less than the true Fullness you are!" I hear Jesus saying move from "less than" to I AM Consciousness. I understand that Jesus is one of our ultimate avatar examples of Consciousness Itself. I hear our Consciousness asking us, "Do you want to manifest *God* Presence or *God* Absence? It's entirely up to you! *God* Presence is already here. If you want the perception of *God* Absence so it is." In our enlightened state evolving each moment I'm quite sure our answer is, "I think not! Let me try again!"

This line of thinking is why the Abraham teachings are so important to me. To define anything as not *God* is simply our grand illusion of our separateness; what the Eastern mystics have been saying for millenia. As many have said before me, if I feel separate from *God*, *God* isn't the One who moved. **My** awareness moved to a different point! *God* is Spirit. *God* is everpresent. Our human projection of judgment makes it appear to us that *God* has removed Itself from us. I categorically know that is not the nature of *God*. As the

Abraham teachings remind us we need contrast in order to see in a physical Universe. We don't need to be afraid. We can trust Nonresistance to what is. Same message, different wording: Jesus reminds us: We are not to judge. We are simply to love and appreciate what is, as is! If we don't like what we see align with Love. It's already here!

As we consider what *God* is it is important to understand that by defining *God* as we do we're also defining our understanding of our relationship to Power. Isn't it obvious that the old definition defines Power as coming from outside of us? I have always intuitively known that Power is within us!!! That's just one reason why I love Goddess spirituality so much. Goddess spirituality affirms that Power is immanent. Inside us! Not outside of us! Divinity is immanent, therefore, Power is immanent. Goddess spirituality is based on the principles of "Power within" and "Power with", recognizing the equal value of **every** person, honoring shared equal leadership, and community, rather than the patriarchal values of "power over", competition, and dominance. I admire Rev. Dr. Matthew Fox so much for his passionate devotion to Goddess spirituality and the feminist teachings that speak to the overarching need for change in the church. Matthew Fox is so devoted to equal opportunity, equal voice, equal power for women in every capacity in the church. Matthew Fox has endured great personal hardship for his views. It is important to me to acknowledge that Rev. Dr. Matthew Fox is just one of a growing legion of white men who acknowledges that the power structure of the patriarchy has to be dismantled. That takes great courage. I am grateful for anyone who shows this level of courage and rises above the personal sacrifice that often comes with "speaking Truth to power." It's so important to speak up any time change needs to happen!

At issue, of course, is the old definition: the judgmental, bearded white man in the sky image. By the old definition we're in an unequal relationship where something outside of us is going to reward us or punish us for our deeds. Here's where the rubber meets the road, with a rather loud thud, and I really love it! Toni Morrison, Pulitzer and Nobel prize winning author, said, "Definitions belong to the definer, not the defined." Her statement echoes clearly for me the understanding that we define ourselves by what we think, say, and believe! We may think we're defining "other" when in actuality we're defining ourselves and our world!

One old defining point that I so very much want to bring a new understanding to is the age old request, "I want *God* to intervene and have so and so do such and such." This disempowered stance completely misses the point that *God* is immanent and everpresent. There is no need for intervention! *God* is already and always present! **We** need to be present to this Truth! This age old attitude reminds me of several quotes I love: "*God* can't do for us what we're unwilling to do for ourselves." The other quote is, "*God* helps those who help themselves." These quotes speak to my new definition of prayer.

For me, the purpose of prayer, the very essence of prayer, is to align ourselves with the Good we are; the Presence that's always present in us, **as** us!!! It's a gross misunderstanding of prayer to think that we can control someone else by demanding that they change while we stay in an unawakened state with the essence of our being. Like saying, "I'd be fine if so and so would just change what they're doing." That's not prayer. It's manipulation. That's an example of "power over" which effectively disempowers both people, and any society that functions in that dynamic. Coming from that stance we've completely missed the point that Power, *God*, is within us, and that there are changes **we** need to make!

Real change comes from within by realizing we have created our outer world by what we're doing with our inner thoughts, words, feelings, actions, beliefs. I find it rather comical and yet oh so powerful that the moment I change, the "problem" no longer exists! Getting this point is the key to happiness and joy of every kind. In order to reach our full potential I so very much see that we need to release this old definition of *God* and embrace the *God* Presence image of ourselves as one with this infinitely Loving Presence within us, **as** us, all around us. I see Toni Morrison saying to us be clear that someone else's definition of you doesn't define you! Consciously defining ourselves is an active process, not a default process! For our highest holiest good to be made visible we need to be acting in our lives as though we know we are powerful! It is no accident that we are who we are! We are Life Itself in process!!!! Enjoy It!!!!!!!!

This is just the tip of the iceberg of all icebergs. Don't let someone else's old worn out definition of *God* define what *God* is for you, or what *God* is for all of us! It's time to step into the Cosmic Void Itself and feel the radiant resonance of this new definition. I know I'm certainly more than ready for this new definition to be understood and embraced worldwide. I feel so much Joy, Love, and Peace in this new definition! I feel such a welcoming Presence, as though *God* within us is saying to us, "Welcome, I have been waiting for you to wake up to who you are!"

We **are** recognizing ourselves and we are coming into our own! Quantum physics is now showing us in physical reality what the mystics have been saying for years. The Cosmic Void is anything but empty. The Cosmic Void contains all there is, all there ever will be all at once. It's all present in the Cosmic Void now! We are one with all of Life, all of Creation. Quantum physics experiments are showing us that one atom affects another atom even when the other atom is contained in a completely lead-cased enclosure at a different location. When one atom moves the other atom moves to meet it.

All of this is saying to me that *God*, Cosmic Consciousness, is **so** much vaster than we've ever dreamed. We are far more than just a physical reality. Our understanding is expanding because our dreams are getting bigger. We're waking up to the vastness of what *God* truly is, and we're growing with It. All the definitions of *God* as Light, Energy, Essence, Presence are all true. That is what we are as well! We are Light. We are Energy. We are Presence. We are the living embodiment of *God* made visible. How much more exciting and inviting can Life possibly get? I'm so very grateful for this awakening awareness in our humanity.

My extension of Toni Morrison's statement is my own belief in the power of one, and my belief that we're all here together! It takes one person to put forth these ideas. What I love about our spiritual development process is how the principles of synchronicity and critical mass work. It is time for a new understanding of what *God* is and that time has come. I feel we've reached critical mass worldwide and the world community is saying, "I want more Love and Peace in my life." This growing understanding is spreading rapidly.

So many spiritual traditions honor the principles of "power within" and "power with" rather than the old, outmoded fear-based "power over." The fear-based model is dying out. It may not seem like it with all the wars going on but I feel the wars are asking us directly and clearly, "Is this what you want as your model of *God* in the world?" Our answer is resoundingly, "No! I want Peace. I want Love. I will work toward living It all into being here and now! I trust in my ability to live in this awakened awareness for all humanity."

I feel this move toward Love and Peace is evidenced in the worldwide resurgence of the Goddess traditions and the increasing visibility of the spiritual teachings of the New Thought movement. Both of these holistic spiritualities honor:

the immanence of Divinity, the sanctity of all life, mutual respect for all, the inherent worth and dignity of all life, diversity, community, cooperation as just some of their many shared values. I love these values. These holistic values express this expanded definition of Divinity that quantum physics is showing us. I love that because it expresses the Generosity, Magnanimity, and Joy that is the very Heart of the Divine Itself in us and all of life.

It's not lost on me that people are struggling with this move. People are hanging onto the old ideas, the old understandings with all their might. They're afraid! They don't understand anything else but their old way of being, doing, believing, thinking, feeling! Sometimes I'm not patient with the fear and the ignorance. I have to stop and remind myself of the expanded role I want to play in all of this. This is where the Abraham teachings about nonresistance, contrast, parallel reality are so important to me.

I am amused and empowered by the idea of the need for contrast so I can see what I truly want to be. Rather than a judgmental, fear-based reactiveness I can now more clearly, simply say, "Thank you, Universe within me, for the contrast. I want Peace." I understand more and more that as we live from the parallel realities of Love, Nonresistance and Peace, all else will fade away. It truly won't be needed! I understand if I want Peace I must be Peace. If I want Love I must be Love. I'm already both of those, and more. I must be them, live **from them**, consistently. It's an inside job. I can't "get It" from others.

I understand that resistance simply makes fear stronger and seem more real. I am like the big sister to my small frightened child within me. I'm saying to this beloved child, "I love you. I will help you. I am with you all the way through this growing process." I will do this for myself and all others! That is my purpose in the world. I believe that is our shared collective purpose in the world. That is why I feel this expanded definition of *God* is so needed and important in our world today. I am here to "be the change I want to see in the world." I join Gandhi's living memory of bringing peace, nonviolence and mutual respect into the world as a living reality for me and for all!

Another aspect of my redefinition of *God* is expressed in the Gaia Hypothesis as put forth by Brian Swimme defining the Universe as Self-reflective, meaning the Universe is aware of Itself in process. I hear in this why Love and Peace are so important to us. I hear in this our recognition of our Source, our Divinity. This understanding is one reason why the Goddess, New Thought, and Gnostic spiritualities are all so important to me. They all share a valuing of our inherent Intelligence, an understanding that we have the inherent Intelligence to recognize *God*. We don't need any intercessory relationship with *God*. We have a direct connection with *God*. We are individualized expressions of *God* Itself.

This understanding speaks directly to my understanding that our definition of *God* is an expression of our relationship to Power. *God* is Power Itself. We are Power Itself. We don't need to go to anything outside ourselves as though we have no power within us, no awareness, no intelligence within us. We are direct expressions of Power, Intelligence, Awareness by virtue of who we are! Realizing this takes fear completely out of the picture! There is no "other" power. There is no devil.

I am reminded of one of my favorite statements from Reader's Digest years ago, "Fear knocked at the door. The door was answered. No one was there." I understand this to mean the only power fear has is the power we give it by our attention to it! Fear disappears when we go at it directly! Saying there is a devil is a misguided understanding of what fear is.

Fear, correctly understood, is a necessary component of physical survival. The emotional, spiritual fear that so many people are paralyzed by is a socialized acceptance of disempowerment. The “power over, God-outside-of-us” dynamic seems so real, so big, people don’t realize there is another choice. We don’t seem to realize that by accepting responsibility for our own lives we empower ourselves. There’s no need for a scapegoat. A scapegoat, such as the devil, just continues to put off our “response-ability”, our empowerment!

I see this disempowerment dynamic so clearly played out in our past. Education and literacy were viewed as only the province of the rich and powerful; for men only, not for women! People were purposely kept poor and ignorant so they wouldn’t and couldn’t upset the power dynamic of the day. The emergence of the printing press blew a hole in that dynamic. People wanted to be educated and they now had a mass medium that made it possible.

People were killed for bringing books to the poor when that wasn’t the socially accepted thing to do! This sociocultural dynamic is an example of how changing our thoughts, feelings and beliefs entirely changes who we are in relationship to power, to ourselves, to each other. Changing our thoughts about ourselves and each other changes our relationship with, and our definition of, *God*! One by one we choose to band together and we echo Dr. Margaret Mead’s understanding: “Never doubt that a small group of thoughtful, committed citizens can change the world, indeed it is truly the only thing that ever has.” That small group starts with one!!!! You and I are examples of the one!!!

Another example of how our definition of *God* plays out in our lives occurs when we consider the Bible as the ultimate book of revelation for Christians. For the past 35 years I have understood the Bible as a metaphysical allegory. I read the

Bible in the context of Dr. Bart Ehrman's statement in his landmark book, "Misquoting Jesus", which serves as the wake up call of wake up calls for me. Reading Dr. Ehrman's book was an eye opening exhilarating experience. His comments only further confirm why I so highly value our own inherent Wisdom that is our very Nature. I view the process of how the Bible came about as a defining moment in the sociocultural story of our humanity.

Dr. Ehrman's key statement for me is, "Not only do we not have the first original, we don't have the copy of the copy of the copy of the copy of the first original!" Then add to that revelation his statement that "in the days before the printing press, the Bible was being hand copied and recopied and distributed by scribes who couldn't even read well enough to copy correctly." Never mind the issue that the scribes of the day may have disagreed with what they were copying so they deliberately changed what they were copying to suit their beliefs. These mistakes got accepted as the correct translation!!!

Then add to that reality that the Bible was written over a period of hundreds of years so the people writing the Bible weren't even alive to hear what Jesus was saying. Add to that how power dynamics were used, and are still used, to frighten people by defining anyone who disagreed as a heretic and any view that didn't agree with the accepted notion as heresy and that gave the "powers that be" the right to kill the offenders to keep everyone safe. I don't need nor want anyone to keep me "safe" in that manner! That's not safety! It's abuse and disrespect! Isn't it the ultimate irony that the very root word for religion means "to link." The Latin root word is "religio", to link!!!

All of these examples serve the purpose of making it very clear to me why I so highly value our inherent Divinity, our inherent Intelligence and Wisdom, our Self-reflective capacity to discern for ourselves what definition of *God* rings true for us. It is so very important to understand that we are either consciously or unconsciously defining what *God* is in our lives by every thought, feeling, word, action, belief we put into being in our lives!

Defining *God* and our relationship with *God* is the ultimate personal wake up call. Understanding that we are *God* in expression is the ultimate gift. I see this process as the ultimate reason we are here. That is why life is so special to me! That is why the work of empowerment that I do is so important to me. We're meant to live lives of exponential Joy, Peace, Love, Wisdom, and Abundance. Knowing who we are makes it possible.

Being Self-reflective spiritual beings also means we are Self-Confident beings! We recognize our *God*Self! We can't be anything but Self-Confident as we radiate the Love, Peace, Joy, and Wisdom of our true being, our true Nature! We recognize ourselves as individualized expressions of the Infinite. We recognize ourselves as Co-Creators with the Infinite Oneness. How much more peace could there be than that realization?

It cannot be overstated that how we define anything or anyone defines how we see them and how we see ourselves in relation to them! Language is powerful. Concepts are powerful. Perceptions are powerful. Language, concepts and perceptions color our world very dramatically. We see this demonstrated every moment in our lives today.

In the 1970s the APA (American Psychological Association) set the guideline that all Psychology textbooks be written in nonsexist language, recognizing that language is powerful! Research so clearly shows what happens to expectations when everything is defined in only male terms. One classic experiment shows this dramatically. A classroom of seven year olds pretended to be the opposite gender. The girls who were pretending to be boys now expressed goals of being pilots, engineers, doctors. One boy pretending to be a girl expressed, "Well, then I guess I'll be a nothing!" I was so stunned hearing that statement and reading the results of this study. In a matter of minutes the expectations of these young children were so dramatically and traumatically influenced. Another experiment, "the blue-eyed experiment", highlights similar results when children are asked to pretend that they have brown eyes instead of blue eyes, when blue eyes is expressed as the "preferred group." Again in a matter of moments the children are visibly traumatized by even pretending to be in the nonpreferred group. It's very important to say that these results aren't one time flukes. These experiments have been replicated (repeated) many, many times, with children and adults, with the same devastating results.

Another example, unintended as I know it is, is that many of the writings that use nonsexist language still always list the male first. This kind of process still indicates secondary status to the female. It is a matter of consciousness raising to change these habits of thought and language. This consciousness raising process is why I do not use the word history, because "his-story" signifies only the male story. It specifically excludes the female contributions to the story of our humanity.

The point is, using divisive language of any kind sets up divisive values, attitudes, thoughts, feelings, beliefs and actions. That's why defining *God* in all inclusive words such as Great Spirit, Animating Principle of All Life, Creative Principle, the All in All, Holy Presence, is so very important to me. These expansive words open up a whole new world of Joy, Love, Empowerment to be lived here and now for everyone. Defining *God* as only male sets up a denigrating duality for everyone! Fear lives in that place! Using all inclusive, nonsexist names offers an expansive, endless, welcoming, open view. This inclusive view opens out into the massive beneficence that I feel characterizes Divinity. I love this view of Divinity. I am included in this view of Divinity. Love lives here! That means everything to me! I believe this inclusive view of *God* will bring Peace to our inner and outer world!

One more aspect of this *God* definition process that is important to me is defining *God* as Mystery. Separation duality's definition of Mystery says we don't know ourselves and we don't have the capacity to know ourselves. I don't believe that and I find that idea demeaning. By our very Nature we know *God*. We are intuitive by nature. We are Spirit in expression. We are the very breath of Life. We are one with the whole of Life. The known is always unfolding in each moment. The more we invite the Presence into our lives the more we know our *God* Self. There is no mystery that each Now moment doesn't know!!

I love the Oneness idea of *God* as Mystery. *God* as Awe and Wonder fits perfectly with the Magnanimity and Grace that defines my very understanding of *God*. I have been awestruck time and again as I watch the Beauty and Majesty of the Universe unfold within me and all around me.

In the early 80s I wrote a poem titled, "Reflections Of A Budding Mystic." I remember feeling odd at the time thinking of myself as a mystic. These many years later I now understand we're all mystics. We are the Mystical in physical form. Many of us just don't consciously know that yet and thus our unhappiness. We feel separate. We are not!!!

In the same time frame I was also teaching a class titled, "Awakening To Our Prayer Consciousness." That was a very fascinating experience. By that time I had a radically different understanding of *God* and prayer than what I was taught as a child. Two of my students were my parents! They were very supportive. I was, and am, very grateful.

In the same time frame I also wrote a paper on the metaphysical interpretation of the Gnostic Gospel of Thomas. I love the Gospel of Thomas. Thomas so clearly expresses the understanding of our Oneness *as* the Divine that means so very much to me.

As we go forward with this expanded, inclusive redefinition of *God* I would like to share with you a little more of who I am and why these ideas are **so** important to me. My background and training is in Psychology. I have a Master's degree in Psychology. I am an educator and a mental health professional. I am an artist and a scientist. I am a published poet and author. I love the gifts of Spirit that a balanced heart and mind bring to my life. As a child I was a tutor for younger children. As a teenager on Senior Day I taught a Latin class in our high school. I will always remember the words the teacher said to me that day: "You'd make a great teacher!" Even then I could see how scared the students were to speak up and take a risk to answer my questions even though I was only three years older than they were. I encouraged them to speak up. I praised them for their attempts to translate their lesson. The teacher acknowledged me for my understanding of the lesson and my ability to relate to the students. I appreciate her understanding of who I was at the time. I am still the quintessential optimist. I have always been interested in knowing the widest, broadest, deepest understanding of all there is to know. I have always rejected any notion that limits our potential and our greatness in any way. I have always supported every notion that propelled us toward our potential and our greatness in every way. I understand that this optimistic, empowered attitude is why I define *God* in the manner that I do. I understand that is why I am who I am today!

In light of this understanding I posed two questions recently to myself and others on my website, www.cathycombs.com. I wonder if anyone is doing research on the correlation between optimists and pessimists and their definitions of *God*. Do optimists define themselves as inherently one with the Divine? Do pessimists define themselves as inherently separate from the Divine? Secondly, I can't help but wonder about our biological basis as humans. Is there a statistically significant difference between how a fear-based person defines their relationship with *God* and how a trusting person defines their relationship with *God*? If we could change a person's biological predisposition toward fear could we change their definition of *God*? Could an individual be taught a more hopeful, uplifted attitude? A friend of mine added a fascinating question to my thought process. He asked, "Are we "hard-wired" for *God*?" I'm quite certain we are!!! I believe we are hard-wired for survival, for Life. *God* Consciousness is our highest, holiest expression of Life. *God* Consciousness is our innate Inner Wisdom, our recognition of our Oneness as *God* Expressed. I believe we are evolutionarily hard-wired for the upward spiral spiritual development that Jesus attained in his lifetime. He fully recognized and actualized his heritage as Spirit made visible. We are to do the very same thing. We are Spirit made visible. It is our destiny to develop our highest spiritual potential. That is why we are here on this sacred Earth which is Spirit made visible!!!

Another question I will expand on throughout this book is how we got to the point where we defined *God* as only male those many millenia ago. I fully realize that Jesus' day was a time of patriarchal culture, but that attitude was intermittently alive for centuries before Jesus' time and the different religious cultures got along peacefully side by side for centuries! I also find it fascinating that Jesus patently rejected the patriarchal culture of his day. He routinely and repeatedly spoke to a different order for the day. I believe he understood that the patriarchal culture was based on fear.

I have always wondered about the biological basis of fear and how fear affects how we define our world. I remember the first time I wondered about physical strength being valued as a male quality and how that value might have gotten translated into the cultural value of "better than" and how that value was related to *God* as male, *God* as Protector; *God* by strength, by force, will keep us alive. All this is based on fear and our lack of awareness of how to survive in a physical world that we view as threatening. We don't understand how the world works so we view it as threatening and we think that some big, powerful Presence outside of us is going to protect us from everyone and everything because we don't know how to protect ourselves.

You see where I'm going with this? This questioning process helps me transform the personal pain I've experienced as a result of the patriarchal values and the patriarchal process that is still so ingrained in our worldview. I want to be very clear too that I believe that patriarchal, "power over" values of any kind hurt not only women and children but men as well. I've said this for years. I'm here for the whole world. Not just for me. I want to be very clear as well that fear is not the entirety of my being, nor is it the entirety of anyone's being. I want to move us toward the values expressed in a few of my favorite quotes. One by Emerson says, "Do not look back in anger or forward in fear but around in awareness." Another quote I love is attributed to St. Francis of Assisi, "There is nothing so strong as gentleness, and nothing so gentle as real strength." These quotes express the definitions of *God* **in us**, **as us**, that I want to live each moment.

These questions I have raised and many more like them are the forefront of our new sciences and what we can ethically and responsibly do to help ourselves as spiritual beings having a human experience. I look forward to this ongoing spiritual evolutionary process guided by inner Love, Peace,

Joy and Wisdom. I look forward to a new definition of fear where we understand, like author Erica Jong, in my 2007 Woman's Calendar titled, "Believing In Ourselves", that "fear tells me I'm going in the right direction." This new definition speaks to "power within" and "power with", viewing fear as excitement, as opening out into all we are, rather than the shut down mechanism of the past! We are co-creations of an Infinite Love! There is nothing in our way except what we put there!

◆ ◆ ◆ ◆ ◆

Imagine my ongoing surprise that more than six months into this writing project the next wave of understandings blows through. *God* is everything ~~ that's what I've said so far. Imagine my surprise when I further realize: *God* is everything because *God* is all there is!!! If we ever see anything else but *God* standing in front of us we are seeing our caked over misconceptions, misperceptions and projections!!! I finally do have an embodied understanding of this Truth. It is astonishing and liberating. Just like the mystics have been saying for millenia. I finally get it! How perfect still that what I said in this next chapter that I wrote months ago still flows perfectly into what I just added in this ending paragraph of chapter 1.

◆ ◆ ◆ ◆ ◆

Study Guide Questions for Chapter 1

Before I lay out questions for you I would like to share with you what this whole process means to me. I am **so exponentially happy** as I continue to step into this process. My highest holiest intention in this process is **knowing** *God* is everything! The subtitle of that knowingness is: There is no such thing as a stranger!!! That is the: Love, Calmness, Confidence, Generosity, Joy and Faith I want to live my Life from. I know with all my heart we are here to live Peace on Earth now!!! I invite you into this adventure with me.

1) If this definition of *God* and prayer is new to you what does it feel like to enter into this new domain? This is a Life-changing process! What does that feel like? Your Life is probably already shifting. What does that feel like?

2) If this definition of *God* and prayer is not new to you you're still likely to meet many new ideas and feelings. What does that feel like to you? How is your Life shifting as you step further into this new territory?

3) What does it feel like to embody the realization that you are Love Incarnate? What old tapes (old learnings) do you have to give up to allow Love Incarnate to live as you?

4) Perhaps this book and question process rock the very foundation upon which you built your Life. Can you allow this new birth without blame and judgment? Will you allow room for Love, Joy, and Peace to live into being **as you**, like never before?

5) What does it feel like to have a different definition of, and relationship with, fear?

6) What does it feel like to have a different definition of, and relationship with, Power?

Chapter 2
What Are We?

I see us as co-creations of the Infinite and co-creators with the Infinite. I say co-creation because I understand that creation is not a one-sided process! I see us as unlimited spiritual beings. The Infinite is unlimited, therefore, we are unlimited. I see us as Consciousness Itself. Charles Fillmore, cofounder of Unity with his wife, Myrtle Fillmore, defines consciousness as "the sum total of our thoughts, feelings, beliefs and ideas."

Myrtle Fillmore exemplified this understanding by healing her body of tuberculosis, which in the 1800s was accepted as fatal. Myrtle did not accept that tuberculosis was fatal. She saw herself as co-creator with *God*. She knew healing was not only possible but inevitable as she held to the Truth of her being. She was completely healed through the power of her faith in her *God* Oneness realization.

This is just one example of how our life dramatically outpictures our consciousness each moment. It's no coincidence. No accident! What we're thinking, feeling, believing, and imaging manifests in the outer! This is the spiritual law of Cause and Effect, or the Law of Mind Action.

I see prayer as an expression of this definition. I see prayer as a statement of our consciousness. I see prayer as a statement of our identity. This is important because our prayers tell us whether we are focusing on *God* Presence or *God* Absence! I find it to be so heartbreaking that anyone ever taught the begging, pleading approach to prayer. This method of "prayer" is the epitome of what I call focusing on *God* Absence. I capitalize Absence because it defines the very identity of *God* when we focus on this as our relationship with *God*.

I use the term Quantum Prayer to express a completely different relationship with *God*. Quantum Prayer expresses my understanding that we are Spirit made visible. We are *God* Essence made visible. Quantum Prayer acknowledges our oneness with the Infinite Presence! Quantum Prayer expresses our beingness **as** the Infinite Presence. Quantum Prayer acknowledges our worthiness. Quantum Prayer acknowledges that Spirit's only way of being visible in the physical world is *through* us. As we actualize that awareness there will be no more war!

How can there be war when we actualize that we are Spirit Incarnate? Sustained war is incongruent with Awareness! War is a temporary state of being asking us, "Is there something you need to readjust?" That's the difference between holism (oneness) and dualism (separation). Oneness is Spirit. It is permanent. Dualism is temporary. Separation disappears as we realign with the Infinite Presence we are.

What is particularly **huge** for me right now in answering what we are is the issue of Nonresistance as a spiritual awakening process and tool. Number one clue: Identity is **not** linked to outcome!!! If we don't embody that realization there will always be resistance because fear is **so** present!

The other day I heard myself ask myself something like, "What makes you happy?" I was astonished to hear myself say, "To get my own way." I understand that to mean several things. I understand I was so totally disconnected from *God* Oneness I thought my good comes from outside of me. I also understand I was so identified with pain in that moment that I forgot the precious awareness, "All things are working together for my good and I am working with them in the Love and Wisdom and Power of Spirit."

Many months later I further understand my "Getting my own way makes me happy" statement means, "I'm at the limit of my problem solving capacity." I'm feeling, "That's all I know." I've completely forgotten Spirit even exists in me. I'm so identified with pain as the entirety of my being I feel no room and I leave no room for Generosity. Even though I shared some of these thoughts in the above paragraph written many months ago I now have a much more embodied understanding of Generosity. I've moved out of pain identified seeing and being! This little Ouruborus tale and tail is yet another reminder I've come home to my Self. I've remembered one of Nonresistance's key names: Generosity. I've "re-membered", re-embodied: Trust! Peace! Wholeness! Expansiveness! Oneness!

Recently I heard a powerful and precious reminder of our relationship with Life. Nonresistance is our willingness to say, "Thank you, *God*, for the lesson. Thank you, *God*, for the blessing!" This thought was shared with us by Rev. Sallye Taylor, a Unity minister I greatly admire, in her Sunday lesson with us. I had to laugh when I heard Rev. Sallye say to us, "Don't you know that you get mad at *God* when you don't get your own way?" Rev. Sallye also mentioned our relationship with *God* as "tit for tat." Beyond the bargaining I hear so well in that statement I hear the evenness that we expect which speaks to me of our identification with lack and scarcity rather than Fullness! We're saying, "I will give this minimum and You better come up with Your end of the bargain." That statement so clearly expresses the feelings of frustration I was having the last few days. I was so clearly not identified with Fullness!

Rev. Sallye also said to us, "When your money is funny you can be sure you need to forgive someone." She was speaking to us of the need for forgiveness so we can be in the flow of our blessings. Rev. Sallye shared with us the story of Joseph and his brothers. What his brothers meant for evil, *God* meant for good! Rev. Sallye spoke of Unity's definition of forgiveness as, "Give Love for hurt!" I understand that statement to mean,

"Give Fullness for emptiness. Give Generosity and Awareness in each moment. There is no lack!!!!"

Bob Burg and John David Mann, co-authors of the book, "The Go-Giver", echo this view of Generosity and Awareness. I have read their book four times so far!!! I will be reading it time and again!!! They say, "When we focus on getting rather than giving we have the process backwards." We're saying, "Give me fire first and then I'll throw on some logs."

We don't understand Life is an "inside out" process, **not** an "outside in" process. There is nothing on the outside that isn't on the inside first!!! This is the difference between Oneness (everything is already within us) and duality (everything comes from the outside).

We **must** understand which worldview we are choosing. It's an interesting paradox. We **can't** receive when we think we're empty. We don't or won't open to receiving. When we need the very most to receive we say, "No." I've seen this dynamic played out time and again. It's sad and it feels awful. As always it is an opening for awareness if we allow it to be.

I think one of the most profound mistakes we ever make is to believe that pain is a punishment. That thought process comes from our belief that we are inherently evil, we are "sinners", and therefore, separate from the *God* of our being and that we deserve to be punished. That thought process comes out of our misunderstanding of duality, and our misunderstanding of who and what we are!

In this vein I want to share one of the most astounding and meaningful comments I've ever heard. In the Abraham teachings we are reminded, "Be aware that rage is a movement toward relief. It is a movement up the ladder of vibration. Rage is a movement away from despair. It is a movement toward

Love and Joy!" This process is a hauntingly powerful reminder to me of a statement I made some 30+ years ago, "Beware of the pendulum as it swings from psychopathic rage to the "doormat mentality." There is a middle ground called Love."

I love all of these precious reminders of how to live in, and from, the Flow! These teachings of Divine Love, Nonresistance, Generosity and Awareness that I love so very much are prominent in all of the spiritual traditions I love. Every moment is a reminder of what I am and what you are! We are spiritual beings having a human experience.

There is no adversary! Divine Love is always present! *God* is always present in us, as us, through us, for us! Pain may be part of our experience. It is not the entirety of our experience. When we move through the experience, rather than identify with it and get stuck in it, we can show ourselves and others the way out through the other side!

We have, therefore, extracted the blessing in our experience. Like mining for gold we have recognized the gift and we have released that which we no longer need. We don't carry any excess baggage! We keep only what is useful, radiant and joy-filled! I really know that understanding, appreciating and living this lesson is the difference between suicide, or embitteredness, or transformation and resurrection! It is the difference between Fullness or half empty or disconnected.

God is still all of this! It is just a question of what definition and what experience of *God* do we want? We can have a radiantly alive, joyful, empowered definition and experience or a crushingly small, dead definition and experience! It's our choice! It's our "response-ability" now and always!!!!!!!

This awareness feeds directly into another affirmative prayer I've said thousands and thousands of times in the work I do: "All things are possible with *God*." I understand that to mean when I focus on my oneness with *God*, as *God* working through me as me, all things are possible. It is my responsibility to align myself with Oneness.

I no longer use that wording in the work I do because I realize most people hear that as *God* is still outside of them, that they are to remain passive and *God* will do something outside of what they are doing. Nothing could be further from what I believe. Having said that I want to honor the fact that I have been recently made aware of an older version of that prayer that says, "When I abide in *God,* all things are possible for me." That is a much more active, co-creative understanding. That **is** what I believe and know to be true!

I want people to know they are powerful. *God* is present in every moment, **as** every moment. ***God* works through us, as us**. I want people to know: By our definition of *God* we are calling our world into being! What is it that we want? What do we think is possible? Look to our definition of *God* for that answer!!!! Perhaps we will know very quickly we need to change our definition and our relationship! All we need is a willingness. The Exponential Universe within us and all around us **is** that magnanimous. All we need is a willingness to truly know the very Essence of our being. The Universe within us, **as us**, will take it from there filling us with ideas to put into action! The name of this Activating Principle is Love! Be conscious. Be willing. That's all we need!

One more thing I must say is I fully recognize there are nearly seven billion people we share this precious planet with, and many of these people categorically reject the understandings I am sharing. The key lesson I am continuing to learn is there is a place for everyone under the Stars, Moon and Sun of our physical Universe. The spiritual gift of Nonresistance is to be able to stand in Peace with anyone who categorically rejects who I am and what I'm saying and to know there is room for Love. We are one in Spirit.

There are simply seven billion, and more, ways to arrive at our understanding and expression of one in Spirit. As I embrace this incomparable diversity I live as Love and Peace! There is no fear. No defense. That is the ultimate expression of Love, Courage and Self-Awareness. It is the ultimate expression of *God* Awareness. I affirmatively desire to live from this place always! Eternal, abiding Peace and Love will be made visible in our physical realm when we do this.

As I say that I cannot help but think of people I admire who are so passionately working toward making Peace a living reality in our world once again. I honor Dennis Kucinich and Marianne Williamson, to name two, who are working to make a State Department of Peace a living reality in our country. I also honor Rev. Rita Marie Johnson, a Unity minister living in Costa Rica, for her efforts to teach Peace to every child in Costa Rica by combining the principles of HeartMath and Nonviolent Communication. Costa Rica is the only country in the world that has made a conscious decision to abolish its army. They made this decision in 1948. What a courageous, forward thinking decision. May we all continue to move in that direction.

◆ ◆ ◆ ◆ ◆

The more and more I listen to myself as I experience the process of writing this book the more and more amazed I am at what I hear and what I remember. I was playing solitaire right now on my computer. I kept winning. I heard myself say, "This is boring." I was immediately shocked at my statement. I remembered the first time I heard someone say, "Peace is boring." I thought, "Oh, my *God*, are you out of your mind?" Just now I had the awareness that somewhere along the line we "bought" the idea that we have to be fighting something in order to be doing something! Yikes!!!!!!! Beyond boring, as ludicrous as it sounds to me, we're even uncomfortable when we're "too" peaceful.

It's *as if* we're waiting for the other shoe to drop because we believe this feeling is too good to be true. That's duality's calling card for "We don't deserve to be peaceful. It's dangerous to be peaceful. It's dangerous to let down our defenses." As I sit here enjoying this wondrously beautiful, peaceful, sunny day I know clearly I do not want to borrow any "trouble." I want to fully appreciate this wondrous sense of peaceful quietness I am experiencing. I can genuinely sit in the quietness and listen to who I am. I can calmly, confidently experience who I am. I don't need to conjure up any difficulty to keep me mindlessly busy from the *God* Presence of this day and my being! I don't need to punish myself. I don't need to earn any reward. I can just rest in the Joy of I AM Presence!

———◆ ◆ ◆ ◆ ◆———

I'm realizing more and more that each one of these chapters could be a book unto itself. The ideas are so huge. Here's where Ouruborus' tail and tale has led me now. One of our key identifications defining who we are, especially in the Western world, is how much money do we have? I hear in a way I've never heard before the unconscious connection we must've learned between having money and problem solving ability. We define ourselves as worthless (do you hear "worth less" than someone else) when we don't have the money we think we should have to be identified as successful.

Instead of Ouruborus' tale continuing to unconsciously eat its tail I feel a new awakeness as I say: **NOT**!!!!!!! I know it is so easy to intellectually know the untruth of this thinking process. Embodying the Truth of worthiness whether we have a penny or not is another thing. In the process of writing this book and doing other projects these last sixteen months I have watched the process of knowing that having money is a security blanket. Having money or not is a visible process. Having the Consciousness of Abundance can be an invisible process. I say can be! It doesn't have to be! I have experienced **so** many times of not having a penny to my name and yet knowing I am fine. All is well. All is well is an important I AM statement. Feeling all is well is a Spirit Identified Ego identity, not the small ego self of separation from *God* Identity.

In a physical world we do have needs. We do have bills. We also have, and **are**, the *God* Identity Wisdom to meet all needs in perfect peace and order. We are Spirit Personified. Living from that place of knowingness embodies the Generosity of who we are now and always!!!! From that Spirit Identified place I find Life to be a fascinating creative problem solving process. I don't mean "problem" in the negative sense we so often see that word. I mean not only possibility but inevitability!!!

I am reminded of a Psychology class I took in graduate school on Creative Problem Solving. We were given a problem (I don't even remember what it was). Our task was to write down 300 solutions to the problem. At first, I thought, "Are you out of your mind?" I sat with the impossibility of that assignment for a moment and then I gave that up. Then I was hot on the trail of my 300 solutions. I found more than 300!!!!

I had answered for myself my basic question, "What am I?" I am Spirit in physical form. Where is the solution? Within me always! I had Self-Identified as Co-Creator. Infinite possibility now and always! The solution is always within me! Nothing is impossible when we identify with Spirit, as Spirit... Our one and only Source. You are the very same. Co-Creator. Infinite possibility now and always! We are Spirit in visible form!

Study Guide Questions for Chapter 2

1) You are a co-creator of your entire Life. Is that a new reality? What does it feel like to live into that reality?

2) To define yourself as the Infinite Presence made visible, what does that feel like? Is that the height of heresy, a terrifying responsibility, or a freeing, uplifting reality?

3) Do you sense a new embodied understanding of: Faith, Confidence, and Nonresistance as you allow room for this process to unfold in you? Are you feeling more like shutting down instead? What old beliefs are being challenged? Are you feeling too visible?

4) How do you see living from Nonresistance serving you or frightening you?

5) How do you define yourself? Is your definition of yourself being challenged as you read this book? How are you working with the challenge if you feel one?

6) Describe your relationship with giving and receiving. What does it say about you?

Chapter 3
Oneness Or Duality? It's Your Life!

My first title for this chapter was "Oneness Or Duality? It's Your Choice!" I do believe our worldview is a choice. I believe it's far more than that too! It truly is our life! The importance of this statement can **never** be underestimated. The **entirety** of our being rests on whether we view the world from oneness or duality. Test that out for a moment if you will. Watch your thoughts and pay attention to the feeling that accompanies your thoughts. Do you overridingly believe the world is benevolent or dangerous? I wonder if you're surprised by what you've just observed.

Our choices, moment by moment, create our life. Most of us weren't taught this very vital reality. To me, even an unconscious belief in oneness sets up a magnanimity and beneficence that is beneficial to every level of our being; whereas, a wrongheaded belief in duality constantly sets us up in a fear mode. This is really important.

I believe duality was originally simply a necessary distinction to be made in a physical universe. Night and day *are* simply different expressions of the same reality; the progression of a night into day in a 24 hour period. Somewhere along the way in our human story *one* person decided to equate night with evil and day with good! Then we, as a collective ***un***consciousness, a collective society, compounded the error by building a whole system of thought, an entire worldview, on this idea! This way of viewing duality is based on fear, separation and domination! It's perpetuated as a worldwide unconscious habit pattern. It's what we learned and never questioned, much to our detriment. This system of being is based on not understanding our oneness with Divinity and each other. We see the implications clearly today **in every area** of our lives throughout the world.

Duality, as separation, is an unnatural state of being! This understanding of duality is an artificially orchestrated habit pattern of thinking, being, feeling, acting. Research consistently shows that constantly being in a "fight or flight" mode is detrimental to our health and well-being. In the Western world we have astronomical cancer rates and heart disease rates for a reason! We can change that *today*!

Today we understand that our body~mind~Spirit oneness does indeed form a connected feedback loop. What happens in our body affects our mind, our brain, our emotions. What happens in our emotions affects our body, our brain, our mind. One example of the importance of this connection has huge implications for every idea in this book. It is so astonishing to experience when an awareness becomes embodied!

Recently, like a flash out of the blue, I realized that people are actually physically deaf when they are emotionally stressed! I sincerely hope research on this process is being done in places like the HeartMath Institute in California. This stress-induced physical deafness explains why I can be talking with someone in person or on the phone and directly answer their question and their response indicates they didn't hear a word I said!

This can be irritating. We might assume, "They're not listening to me. They don't care about me." While this may be true to some extent, I believe something far more likely is happening. I believe we unconsciously disconnect. We dissociate when we're stressed and we don't even realize it. Our body~mind~Spirit balance is in conflict. As a self-protective mechanism we dissociate when we don't have a more appropriate social response mechanism ready.

This is an example of the holism of our feedback loop in process as the body~mind~Spirit connection plays out.

Body~mind~Spirit --- we are one! Our bodies are built to experience homeostasis, a proactive balance. Homeostasis is not stagnant! Stagnation is an inert death process. Homeostasis is a Life-sustaining balance. Homeostasis is a relaxed proactive attention! Homeostasis is the body's expression of oneness with its whole self.

———— ◆ ◆ ◆ ◆ ◆————

Months later as I've further embodied the impact of these ideas I realized that the belief in duality has systematically taught us **to be** personality disorders! I was stunned when that idea came to me.

This is important! Personality is our basic sense of ourselves. Our personality is formed in the first few years of our lives as we begin to understand who we are and what our relationship to the world is. A healthy personality expresses as an integrated whole. An unhealthy personality is dis-integrated and disintegrative, meaning very fragile, brittle, inflexible, not well connected or held together. We don't have a consistent view of who we are in our world. The world seems chaotic and we are scared.

There's a wide range of expression of personality disorders. A full blown personality disorder is a severe mental illness characterized by extreme rigidity in thought and emotional expression, extreme anxiety and fearfulness, and extreme approach-avoidance behaviors.

Perhaps the most important word here is rigidity. Rigidity means an absence of flexibility. Rigidity negatively impacts the compassionate give and take so necessary for healthy happy relationships with our world. Rigidity expresses as oversensitivity to criticism, very low tolerance for frustration, an explosive temper, not honoring boundary issues, not understanding what is my responsibility and what is your responsibility, very low self-esteem. These are personal, psychological, social, political and spiritual issues. We cannot have a healthy, peaceful world community without healing these issues.

In this context of oneness and duality I think it is **so** important to understand ego. From a oneness perspective, ego is simply our individualized personality in expression, our individualized expression of *God* Mind. There is truly only one mind, *God* Mind. That revelation exploded into my awareness recently.

Where we get into trouble with our understanding of ego as separate, as duality, is when we forget where *God* is. *God* is **in** us, **as** us. We are all different expressions of *God* Mind. When we see difference as fear we revert to "power over" dynamics thinking we have to control something or someone in order to be safe. This separation process starts very early in our lives. This separation process starts unconsciously and then becomes a reinforced habit pattern of how we see and approach our world. We see only that which reinforces our view of ourselves and our world and we disregard the rest!!!

This is hugely important. Do you see how we set ourselves up for a self-fulfilling prophecy? When we see difference as exciting we open to a sharing modality of, "Greetings, I want to get to know you better." We recognize a newfound friend. We recognize our oneness! We are not afraid of parts of ourselves that we do not yet know! We recognize all as *God* in expression! We maintain our integrity, our interiority. We know we are safe! Whether we are in known or unknown territory we approach life from what Rev. Dr. Michael Beckwith calls a "reverential alertness." I love that phrase. It expresses the quiet, calm, Joy-filled confidence that I love so very much.

In that same light, I love the definition of ego I just read in my 2008 Mystery School work at Diana's Grove: ".. without healthy self-esteem we become lifeless and we lose our creative energy. Through our egos we touch external reality. With a balanced ego, we become conscious co-creators of that reality." That's the holistic view I love **so** much!!!!!

It is exceedingly important to understand ego for what it is and what it isn't! It isn't **just** separateness as we have come to see it identified in our modern way of thinking and defining. Ego is the executive function of our mind and personality. Ego is an absolutely necessary, appropriate aspect of our whole selves. To identify ego as **only** self-pride (as though self-pride is a bad thing) creates an overly passive, fearful, submissive, shrinking, anxious, angry, disengaged, disembodied, disempowered person, totally out of touch with who we truly are! Ego (our Individualized *God* Identity) is our fully embodied holy preciousness!!! Living from this Identity we revel in our existence. We are openhearted, funloving, engaging, expressive, trusting, loving, creative, giving, generous, expansive, calm, confident, flexible, resilient, filled with awe and reverence: doing our perfect work in the world! The bottom line again is: understand what duality is and what it isn't! We cannot be of service, in calm relationship, with the world when we don't know who we are!!! We must be awake to who we truly are!!! Our *God* Identity Ego **is** awake!!!

I love what Rachel Naomi Remen, MD, one of my favorite people, says about this relationship. This quote is taken from our 2008 lesson at Diana's Grove Mystery School. Dr. Remen says, "Fixing and helping may often be the work of ego, and service the work of the soul." My understanding of these words is: When our "helping" is in the context of: "Get rid of your pain. It makes mine too obvious", that's what I call "small" ego; the agonized sense of separation from Self. My definition of service comes from the attitude: "I love myself. I love you.

I see, and revel in, our connectedness. I see our wholeness. I trust us both. May I be of service to you to reflect back to you your exponential shining Greatness, our collective shining Greatness?" That's service!!!!!!! That's a Gift!!!

I see our work in this lifetime to be the conscious process of sharing our Giftedness; serving the Whole; rising up into healing; embodying Oneness and understanding duality for what it really is: simply different expressions of Essence!!!

Another beautiful expression of this awakening process is also from a quote in the 2008 Mystery School work at Diana's Grove, as written by co-founder, Cynthea Jones. The Sun is speaking to us saying, "The things you think you have hidden are just as visible as the things you properly display. When you discover that all you have hidden is still visible, you will be ready to see yourself from My point of view."

The dialogue continues, "What do we do with the things we need to banish? Transform them. That's the only way to really be rid of anything. You can't hide what you don't want. You only think you can ignore that which you have sent away.... There is no such place as away!" This astrological analogy is like *God* saying to us, "See yourself as I see you and know that everything is just fine. You are loved. You are lovable. All is well in the world. Be at peace." Our work in this world is: See ourselves as "response-able." We are capable of being fully conscious in the Now!!! We are Spirit Embodied!!!

◆ ◆ ◆ ◆ ◆

As we continue our journey let's look back for a moment at how we got to where we are today. Four hundred years before Jesus lived, Aristotle understood oneness! His teacher, Plato, did not! Aristotle was not allowed to teach in Plato's academy. Aristotle founded his own Lyceum, the precursor to our university system.

In Jesus' day the issue of oneness and duality played out in this way. The nature of Jesus was a **hotly** debated issue. There was **not** a unified understanding at all on the nature of Jesus. This is where our very future in Christianity was decided. This is where the very definition of *God* was decided. This is why the Gnostic Gospels are **so** important to me. Apostle Thomas understood that he and Jesus were the very same Nature, *God* Expressed. Apostle John did not share that view. He felt he was categorically different than Jesus.

I find it more than a little fascinating that metaphysically Thomas represents Understanding and John represents Love and yet John didn't love himself enough to understand and know who he truly was. Jesus himself said to his apostles (and to us), "Greater things than these shall ye do." Once again the world chose duality. John's view was accepted by the church establishment of the day. Thomas' view was declared heresy. The penalty for heresy was death.

I'm highlighting some of these radical responses to difference to emphasize the importance of Merlin Stone's point that there was a time when people with radically different views lived side by side in peace for *thousands* of years. Peace is possible!!! Peace is a Conscious Choice!!!

I am passionately devoted to Peace and Empowerment. Violence in any form comes about when we are in any way disconnected from the present moment. When we are in any state of being other than the present moment we are in an unresolved state of being. We are at unrest! When our words~feelings~thoughts~actions are synchronized or congruent we are at Peace. For me, Peace is a reflection of what I consider our highest holiest call: Compassion and Gratitude. There is no violence in this state of being. I do fully recognize that sometimes that call seems like climbing the Himalayas backwards but I know Peace and Joy are our calling nonetheless.

Let's fast forward now to the 1500s to see how this nature of *God* issue was still being played out. The question still is, "Will it be Oneness or duality?" Yet again it's not a smooth ride. How power is used and misused cannot be missed in the choice of worldviews. This issue is huge because yet again it speaks to our definition of *God*. Is *God* an immanent Presence or an Entity outside of us? Is religion going to be a "power over" process or is it going to be an honoring of "power within" and "power with"?

In the story of our humanity horrible things have been done in the name of *God* and religion. That's just one reason I make a distinction between religion and spirituality.

Here's how it played out in the 1500s. John Calvin was so incensed by Michael Servetus' discourse on Unitarianism that John Calvin had Michael Servetus, the first Unitarian martyr, burned **alive** at the stake with his books. Calvin's brutality backfired on him. His brutality so incensed the people the whole country revolted.

Fast forward another hundred years to another rendition of dualism. Rene Descartes, the 17th century proponent of dualism, said that the body and mind are separate. Society sided with Descartes.

Fast forward now to 1909 when the Observer Effect was introduced. This discovery set off the world of physics. This discovery directly contradicted the longheld idea that our physical Earth is just a dead mechanistic system upon which we have no impact.

The Observer Effect says that an observer (the mind) *does* have an impact on matter. It changes the relationship between two entities. To this very day the Observer Effect has an ongoing impact in physics, spirituality, psychology, and the ever expanding discipline of holistic medicine.

The spiritual and psychological importance of this discovery process highlights how we savagely attack a new idea because it threatens our rigid status quo. New ideas often challenge our very identity!!! We often equate this emotional challenge with physical survival. Then we gradually move through the process until we matter of factly accept the new truth as though this truth were self-evident from the very beginning, when it was anything but that when it was first introduced.

Yet again how synchronistically hilarious Spirit is. Just as I was writing this paragraph my phone rings and it's Rev. Mary Manin Morrissey with her "Miracle Minute." Mary quoted from one of my very favorite books, "Illusions", by Richard Bach, "Argue for your limitations and sure enough they're yours." How passionately we have argued for separation and duality.

This rigidity~flexibility process has huge implications for all of us!!! It doesn't matter how brilliant we are, what our field of study is, or anything else. This rigidity of refusing new information because it challenges our own ideas has happened and is happening throughout the story of our humanity. This is exceedingly important because it speaks of mindlessness instead of mindfulness. I don't know of any idea on this precious Earth that calls us more to mindfulness than reclaiming the reality of our oneness with all of Life. Quantum physics is now showing us beyond a shadow of a doubt how interconnected we are. We're in for a thrilling and challenging ride. Quantum realities are sure to change us faster than we may be ready for so a prayerful mindfulness is key for all of us.

It is now 2008. The medical and scientific communities still haven't fully embodied the understanding of the body~mind~Spirit oneness paradigm! It's 99 years since the idea of the Observer Effect was introduced. The current research on the zero point energy field phenomenon overwhelmingly substantiates that mind affects matter in profound ways. The zero point energy field research clearly shows that at the subatomic and quantum levels everything is flowing energy; everything is intricately interrelated and constantly sharing energy information and being changed by it!!!

This is important for our discussion because many of our modern day scientists don't understand that prayer is energy! In carefully controlled scientific studies prayer has been shown to have statistically significant impact on outcomes. That means the outcome **isn't** by chance. Prayer is having a direct impact on outcome. These results have been replicated time and again.

These paradoxes continue to exist side by side. Imagine my surprise yet again as my full blown awareness bursts forth fully embodied yet again! I just now saw what this paradox looks like: We live in a world of duality **when we think we do**! We live in a world of "both~and" **when we think we do**! Imagine my further surprise when the subtitle of my book bursts forth just now some 14 months into this writing project: Quantum Reality: It Already Exists! It's hilarious to experience that reality right now! Yet again that's what fully embodied realization is: "Of course, prayer is already answered. Everything already and always exists in Spirit, beyond time!" We simply call it into physical being by our intention, by our attention to it!!! I'm sure I'll happen upon this realization time and again (pun very much intended!!!). Who knows what I'll know by the time this project is done!

Twenty five years ago I wrote a poem titled, "A Multiple Possibility World." My understanding of what I was saying back then continues to expand in ever more phenomenal, comical and amazing ways. Science is now showing us in ever more phenomenal ways that it so very much matters what we think because we usually attach so much feeling to what we think!!! Love is Energy!!! Fear is Energy!!! **Anything** that currently exists is in physical reality because we brought it into being by our intention. No matter what we name it we brought it here by our intention and attention!!!

Lynne McTaggart, an award winning investigative journalist in the fields of science and religion, is doing a project now called "The Intention Experiment." Her project clearly shows the power of intention on subatomic and quantum reality. Her work shows how subatomic levels of energy materialize as the power of intention is focused on them.

I went to her lecture last night. The final point of her lecture was astounding. She detailed that even people on the lowest rung of the poverty ladder do not show stress if they have two things: a strong spiritual belief and a strong support system. They do not feel isolated, therefore, they do not feel stressed.

Her last slide detailed a polar bear and a sled dog playing in the snow. Rather than our preconception that the sled dog would be the polar bear's lunch, the polar bear just wanted to play! The polar bear and the sled dog were hugging each other! I almost cried seeing that slide. It was so beautiful.

She ended her hour long presentation detailing her global project to test whether the power of intention could bring peace even to the most violent, war torn places on the planet. Her work is important because it shows what is happening at a subatomic, even quantum level, how matter changes when we focus attention on it.

Her work brings forth yet again the power of choice. Intention is a choice. Intention is energy. Love is a choice. Love is intention. Love is energy. Fear is intention. Fear is energy. When we consciously or unconsciously focus energy and intention on wellness it materializes! When we consciously or unconsciously focus intention on illness it materializes! This isn't new! It's just being proven in more and more profound and immediate ways. The body~mind~Spirit continuum is indeed very real! Physical reality rests firmly on the foundation of this Unseen Realm.

Our modern day holistic sciences, philosophies and spiritualities are teaching the value and validity of oneness, of holism. Quantum physics and all the new sciences are blowing holes in every scientific understanding we currently have. I'm so excited about quantum physics and all the new sciences and their continuing impact on our understanding of what *God* is and who we are! These modern sciences are showing us how profoundly huge our world is and how near *God* Essence is!

I'm overjoyed by all I'm learning. In a class I've taken with Dr. Garland Landrith on the EFT process, he recently said, "We're knocking at the door of *God* and *God* is letting us in." As much as I love that statement because it implies such excitement in our newfound awareness of who and what we are and what *God* is, I would say *God* isn't the One keeping us out! We have moved to a new level of awareness of who and what we are. We're creating room for new awarenesses of the magnitude of our oneness.

Through the study of spiral dynamics we see the evolution of our human consciousness. We are once again coming into the original understanding of duality as simply two different forms of the same Essence. This is huge!!! We can see difference without making a value judgment! We recognize we **always** have choices. We are Self-Conscious and Self-reflective beings. We are empowered and connected. We can correct our unconscious choices that were made in fear-based reactiveness. We can let go again and again of "less than" consciousness.

This is an example of the spiritual power of forgiveness. I've heard forgiveness defined as "giving for." I understand this to mean: I release my "less than" understanding for a new awareness of my identity. I recognize that as humans we do make mistakes. Making a mistake does not define the entirety of our being as inherently evil. As co-creations of the Infinite we are Divine in nature, not inherently evil. We are not sinners! I cherish Rev. Dr. Matthew Fox's understanding of who we are; that we came into this world as original blessing, not original sin! Original blessing recognizes and honors our oneness. Original sin perpetuates the separation myth.

The difference in these two parallel views is enormous. Here is perhaps the central example of our power of choice and the Nonresistance that is paramount as we exercise our power of choice. This is where the rubber meets the road each moment. Even though on the one hand I say illusion will disappear as we focus on oneness I recognize there will always be those among us, me included, who will focus on duality and resistance at times.

I know that pain and unconsciousness are temporarily part of our human experience. I also equally know pain and unconsciousness don't have to be spiritually constricting. Coming from Awareness rather than value judgment about what I am experiencing is key in remaining free of spiritual constriction. The more spiritually evolved I become, the more Self-Confident and Self-Conscious I become, the more nonresistant I become. It works the same way for all of us. Nonresistance is my key that I am centered in *God* Presence Oneness. That's how I want to be here saying and doing what I'm saying and doing. I want to embody a more generous, loving, compassionate, holistic view of who we are! By a nonresistant awareness of these parallel worldviews Peace becomes a larger and larger reality for all of us everywhere. One person at a time Peace is more visible.

Peace can also be visible throughout a whole culture. In this context I feel it is exponentially important to address the function of Goddess spirituality. Goddess spirituality has existed for 30,000 years. The research done by Marija Gimbutas, Riane Eisler, Merlin Stone and many others of the current day is exceedingly important in understanding what Goddess spirituality is. Goddess spirituality is **not** just for women or about women. The Goddesses and Gods of every culture represent the qualities of our humanity. For example, Diana is the Greek Goddess of Love. Poseidon is the God of the oceans. Water represents our emotions. These archetypes of our humanity tell the story of our lives. Goddess spirituality is a holistic community-based tradition focusing on the values of equality, community, cooperation, respect for diversity, recognizing the interconnected web of all life, immanence of Deity, nonhierarchical shared leadership, ongoing cycle of life, sanctity of all life, sacredness of the body and sexuality. Goddess spirituality is a totally different value system than the hierarchical, patriarchal values of domination, inequality, competition, and separation.

In the archaeological research done by Marija Gimbutas, Riane Eisler, Merlin Stone and many others there is no evidence at all of war tools of any kind on the excavation sites of the Goddess worshipping cultures. It is clear that these societies saw Peace as a way of being!

As I am writing this book I am rereading Merlin Stone's landmark book published in 1976, "When God Was A Woman." I want to highlight a few of her ideas because she speaks so clearly to the question, "Will we perpetuate the value of oneness or will we perpetuate the fear-based divisiveness of duality?" Merlin Stone agonized over her growing awareness of how pervasive the male bias was in even the profoundly educated male scientists who were excavating the Goddess sites. Even though the sites were clearly burial sites of females, very important females, queens and priestesses, the male scientists' blind bias still brought forth comments like, "That couldn't possibly be the tomb of a female. There are symbols of strength buried with her."

The comment that stopped me dead in my tracks was when Merlin Stone was sharing her process of awakening to the fact that in the Biblical references to worshipping idols they were speaking of the people who worshipped the Goddess. She spoke of how they even deliberately addressed these figures as males, Ashtoreth as male, rather than who She really was, Astarte, The Great Goddess. As much reading and studying as I've done it never occurred to me that by "idol" they meant The Great Goddess!

In our recent lesson on forgiveness this kind of bias is a perfect example of where I need to give Love for hatred. The hatred toward women that is so blatantly expressed in our early human story is almost more than I can bear. The only greater sadness for me is that this blatant hatred of women is so alive today!

Merlin Stone makes the point that these biases pervade even the most secular aspects of our society to this very today! A question I ask today in all seriousness is how much will we have to grow in recognizing the sanctity of our humanity before we are able to acknowledge the damage we do to ourselves and our human community by the repeated telling of the sexist, homophobic "jokes" that pervade our culture? Is a more telling question that we don't even realize we are a community? We think our only way to protect ourselves is to degrade and kill the "other."

This question is important because the reciprocal relationship between attitude and behavior is so evident. Attitude affects behavior. Behavior affects attitude! All these questions resound in Merlin Stone's book. Years ago Goddess spirituality and culture was nearly wiped out by the Kurgan (Aryan) invaders from the North. Goddess spirituality is *still* viewed by many as a threat to the patriarchal values. Goddess spirituality has been demonized as Satan worship. It is nothing of the sort! Words like witch have been used to terrorize and separate people. It is important to understand that the word witch is derived from the Celtic root word meaning wise one. Witches were, and are, the herbalists, gardeners, farmers, midwives, and healers of the day. Today Goddess spirituality is once again flourishing all over the world.

Goddess spirituality is a central aspect of my spiritual practice. It speaks to me as an empowered human, as an empowered woman. It speaks to me of generosity, joy, sharing and cooperation. Perhaps it is important to say that Goddess spirituality is not just for women or about women. I know many men who are passionately devoted to Goddess spirituality because it is an empowering, integrative, spiritually uplifting, welcoming personal practice and worldview. Goddess spirituality is an Earth-Based spiritual practice that honors and cherishes the fact that we share one world together!

A major symbol of this Earth-Based embodied spiritual practice is the five pointed star recognizing the five elements of Life without which we cannot live: Earth, Air, Fire, Water, and Spirit. These elements acknowledge how intimate our relationship with the Earth is. I believe with all my heart that we are spiritual beings sharing a human experience. An empowered conscious move toward love and respect for ourselves and each other would go a long way in healing our precious, precious world! As an educator and mental health professional this healing, community building process is what rocks my world!

In this rock my world framework I want to acknowledge the New Thought religions, Unity in particular since it is the one I know best. I greatly appreciate the move to nonsexist names for *God*. Unity teaches that *God* is in everyone and everything. That teaching gets clouded in Unity's early days when so much of the writing and naming of *God* was male based referencing. This is exceedingly important because sexist wording continues to lump Unity with other traditions that use *only* male-based wording.

Another aspect of this consciousness raising process is the conscious decision to mention women first as often as we mention men first. To always mention men first continues the perception that men are superior to women. Unintended as this may be in Unity it is important to me to raise this issue. I'm grateful for the change I see in this area too. I appreciate the healing this change promotes. As I have said earlier in this book words are important. Words shape how we see ourselves.

While I am critiquing a movement and its teachings I dearly love I want to acknowledge that I recognize these teachings began in the late 19th century and were far ahead of their time. I **greatly** appreciate the modern day Unity ministers and practitioners who are speaking of *God* as not just in us but **as us**, as I have been throughout this book. I think this awareness is so very necessary for our ongoing spiritual development.

——— ◆ ◆ ◆ ◆ ◆ ———

How fitting that I come back to this chapter weeks later seeing the comments I wrote in the previous paragraphs. I am comically aware that this book might seem like a back and forth flip flop between past and present. That's exactly what it is. My awarenesses keep growing as I write this book. The awareness I just had that exploded into my full blown consciousness is that the wrongheaded understanding of duality is a death knell to our sense of Self-worth and thus to Empowerment! Duality says, "You are nothing! You are empty!! You have no value! Anything of value comes from outside of you. Anything of value **is** outside of you!"

No wonder people are so heartbroken. No wonder people live lives of such screaming desperation. I have to say I don't agree that we live "lives of quiet desperation." I feel people are screaming in agony to know who they are, to feel valued, to know they're valued and valuable.

No wonder I am **so** passionately devoted to Self-Empowerment for all! I **know** I am valuable and valued, despite all the messages to the contrary I've received and believed and released over the course of my lifetime. That is why it is so agonizing to me to watch and listen to people suffer. It's unnecessary! We **are** valued and valuable. We are **exceedingly** powerful! We **do** have the Power to live extraordinary lives! We are Power Itself!!! That is our heritage as **beloved** creations of the Infinite.

What further just unfolded into my conscious awareness is Buddha's description of His very name as meaning "Awake." Not until we awaken to our inherent Buddha nature, our *God* Self Incarnate, can we live lives of impassioned glory and significance and meaning. People inherently recognize when they believe they are empty and worthless. That's why nothing from the outer ever fills them! That's why endless genuine compliments go unrecognized and unappreciated. We don't let them in! They feel contrary to what we believe about ourselves and we instinctively know it! We maintain an agonized homeostasis of identifying with our supposed nothingness! Not until we align with Oneness, with our inherent Goodness, are we awake!!! When we consistently identify with our inherent Goodness, with our Beloved Presence, our lives will explode with untolled, countless blessings!!! It can be overwhelming. We're not used to it *yet*!!!

What is **hugely** holding my attention right now is an understanding that I recently embodied like never before: Identity is **not** linked to outcome!!!!!!!!! Separation duality thinking absolutely says exactly the opposite. I was watching my reaction to some events in my life and I finally deeply understand the pain I **so** often feel. I have so often said to friends that I feel like my efforts make no difference at all!!!!! No matter where I mention this thought in this book it has immeasurable relevance to everything I'm saying. I realize now this belief is just another slambang way of believing I have to get my own way in order to be valued by me and the world I think I live in and the values this world holds.

Oneness knows nothing of this devalued "get my own way" mindset. Oneness says, "I love you as is. You don't have to do a thing to earn worthiness. You are inherently worthy. You are inherently perfect." I want to keep allowing space for that Truth to sink into my awareness. I want to keep allowing space for the Stillness that embodies this **hugely** Life-sustaining awareness.

I'm realizing another curveball here is expectation! Expectation says, "I wanted and needed a different outcome than what I got." That is also "duality speak." When I'm locked in expectation I am separated from my Empowerment which always comes from within me. When I am filled with Joy, which is what Empowerment is, there is no room nor need for expectation. I know I am fully capable of filling every need myself. There probably isn't even a need. There is a want and I fill it. There is no concern with outcome. There is no doubt about Fulfillment! Separation ego is *always* concerned with outcome! In separation ego outcome is proof of worthiness! I consider this a major watchdog process for my spiritual development. It's not just an individual process. It's universal. I believe we all share this awareness process.

I want to be very clear here again that I'm not vilifying duality. That would defeat the whole awareness~nonresistance process. I want to keep leaving more and more room for the glaciers to slide over each other more easily. I want to cultivate more and more room for Observer Self to be running the show. I'm fascinated by my awareness of "both~and" perspectives. I'm fascinated by my "multiple possibility world" awareness. I'm fascinated by my awareness that it's all *God*. The pain I wish I didn't feel. All I don't presently understand. It's all *God*!!!!!! It's all Energy. *God* truly is all there is!!!!!!! I'm still on the journey of embodying this precious Truth. Do you see how the glaciers are moving for me? How are they moving for you? Easily I hope!!!

◆ ◆ ◆ ◆ ◆

Study Guide Questions for Chapter 3

1) Do you live mostly from an either/or place or a "both~and" place as you live your life? How do your worldviews, whatever they are, serve you?

2) When have you seen self-fulfilling prophecy at work in your life? What did you learn?

3) Imagine that you live in a multiple possibility world. Can you imagine that you are 100% confident and 100% scared at the very same time? What does it feel like to hold that paradox and countless others like it?

4) When have you felt or been told you were nothing? What part of you knew that was a lie? How do you honor and nurture that All Knowing Self?

5) What is your highest, holiest hope for your Life that seems so far fetched it feels like fear and yet you know it is yours to be and do? See yourself in that beingness!

6) How will you honor Spirit's inherent call from you to you? Do you understand that is you calling to yourself? Now what? How about step 1? How about step 10?

7) What happens inside you when your most cherished values or beliefs are challenged?

8) When have you lost all Hope and Faith? What changed as you grew through it?

9) What gets you through your inhumanity to yourself or our inhumanity to each other?

10) What role does Forgiveness play in your Life? Do you understand what Forgiveness has to do with Resiliency, Compassion, Flexibility, and Strength?

Chapter 4
Nonresistance~An Expanded View!

The title of this chapter is an unintended double entendre. Nonresistance is so central to this book and this prayer process I knew I needed to give our discussion of nonresistance its own chapter. As I gave it this title I also realized that nonresistance means an expanded view of our place in the world. Different as we may be in human form we are from the same Source. So, it is deliberate that I placed the chapter on nonresistance right after our discussion of duality.

Nonresistance is the spiritual awareness of knowing that *God* is present **as** every moment, every person, every situation! Knowing this there is a profound trust in what is, and an appreciation for what is! There is no need to insist on change so we'll be comfortable. There is simply the space for awareness! There is space for the awareness that *God* is present **as is**!!!

I realize that's such a big statement it might be a stopping point for a moment to truly take in what I'm saying here. I wonder how many of us have truly done the work to be comfortable with Life *as is* when Life is presenting something we don't want! If we're saying we don't want it we're yet again not realizing we're the ones who brought it into being by our unconscious attention to something we viewed as negative. It truly does take enormous Self-Confidence to be comfortable with what is! Breathe in. Let it be.

I've had some stunning realizations in the process of writing this book. My biggest realization so far is that nonresistance is an expression of a belief in parallel universes. I have my experience. You have your own experience. Nonresistance means I am clear that I am I, and you are you!!! Nonresistance is a Spirit-Identified state of beingness that has entirely moved beyond the judgment aspect of duality that believes if I am right then you are wrong!!! Nonresistance recognizes what I call the "both~and" realities of two coexisting equally valid frames of reference. From this Spirit-Identified nonresistance we can nonjudgmentally, peacefully listen to a point of view completely different from our own and not be threatened. Nonresistance is a key aspect of our upward spiral of evolutionarily based spiritual development I spoke of earlier. This nonresistant state of beingness recognizes that our survival **isn't** at stake physically or emotionally just because someone disagrees with us!!! There will be Peace on Earth when we embody this Truth!!!

Resistance means I am unconsciously embedded in your life. In that state of unconsciousness I am no help whatsoever to you or to me. As that awareness came over me I could even feel what nonresistance feels like in this new way. I was at work in situations that I would've ordinarily tightened up in, and instead I just laughed. I could see, "Oh, that's what this new nonresistance feels like!" I could genuinely listen to their story and let it be their story, not mine!

That's what the experience of witnessing **as** *God* is! No matter what is being said or done I see and hear *God*! That's powerful because I also feel the peace and ease in body~mind~Spirit that goes with this attitude. That's what agelessness is! There is no time in this place! Another word for nonresistance that just came through me is Love, the absolute Love of the Divine, *as* the Divine. We are the individualized embodiment of the Divine. Being conscious of that fact allows room for amazing results. There is just clarity and peace. No projection. No mud!

◆ ◆ ◆ ◆ ◆

I have to say here that it's months later since I wrote those last few sentences and I'm just now getting it at even deeper levels what I wrote then. Nonresistance is not only an expression of a belief in parallel universes. Nonresistance is the acceptance of different definitions of *God* in expression. Nonresistance is the spiritual understanding of coequal expressions of Truth. I have my understanding of Truth. You have your understanding of Truth. Yet again we're beyond the dualistic (duelistic?) "If I'm right, you're wrong" phenomenon. That's huge. When we truly get that there will be no wars of any kind! There will just be Self-Confident Self-expression!!!! Do you hear that? Now I see perfectly how this fits into my next paragraph that I wrote months ago. I love the synchronicity of this unfoldment.

Another experience I had recently ties right into this nonresistance awareness process, and most likely precipitated it. I was serving as a chaplain during our recent Ash Wednesday service at the Unity Church I attend. I saw myself leaning way to the right with my left hand extended straight out as though I were defending myself. I didn't understand until later what I was doing. I heard me say to myself, "Rise up into healing. There is no need for fear!" I understood that I was standing up into a new embodiment of Strength, Fearlessness, and Confidence. I was understanding that no matter what is happening I am "big enough", "good enough", "present enough" to handle whatever is happening. I can't help but see the new understanding I've had recently of Confidence as a spiritual quality! Confidence, "con-fide", with Faith in my oneness with the One, **as the One in embodiment**, I am Confidence in expression! I am present to the present moment, free and clear! What a perfect segue to our next chapter!

———◆ ◆ ◆ ◆ ◆———

It's at least six months later since I wrote this chapter up to this point. Before we go on to the next chapter here's what I know now. In this fascinating journey we call Life I've been anything but nonresistant lately. What I see and feel most clearly is how small I keep my world when I am resistant to what is! That means I am an exceedingly harsh judge of myself and others. I heard myself playfully say to myself a moment ago, "My God, Cath, is there anything you haven't pounced on lately?" Good grief, Charlie Brown! WOW! It's been a tough, important ride. I see how much Love, Generosity, Confidence it takes to be nonresistant. The fact that I can see and feel these awarenesses also tells me I'm moving through to the other side again. It also tells me I'm moving into new territory where I'm still thinking I need defendedness, as unproductive as it is, and the supposed shield of judgment, as unproductive as that is too! I'm still responding to old, stuck pain. I've been using spiritual techniques of EFT and positive affirmations to help me move into Clarity and Peace~~ into the NOW, not the past!

What I understand more clearly than ever is I get in trouble emotionally and spiritually when I think appearances **need to be** changed. That simply says, "I'm scared. I'm in pain." It means I'm not clear. It means I don't trust myself to handle what is. It means I don't trust others to handle what is. I'm hooked by pain~~ theirs and mine. I'm thinking, "I'm **supposed to** do something" other than simply nonjudgmentally notice the situation. Nonjudgmental noticing means I'm in the NOW! Nonjudgmental noticing changes the situation instantaneously. I don't have to do anything **to** the outer! It's not there anymore! It's already changed! Noticing "what is" changes what is! From Observer Self then we take the actions we are guided to take. We aren't motivated by fear. We are moved by Awareness, our God-Identified Self.

Do you understand what I'm saying? Nonjudgmental noticing (another name for nonresistance) means we have Self-Identified as Spirit, not separation! That's the answer to every moment no matter what! I know I'm moving to a vastly expanded awareness of how worthy of Love I am! Some of my old self judgments are still active and I'm caught in them (even if I don't know what they are). It's time to completely release them. I'm changed. I'm new. It's time to move to the next level and to know I deserve to be there. Spirit Identified Self already knows that! Separation self hasn't quite caught up. This process works the same way for everyone!

What I further understand about my resistance lately is it means what I call small ego has been screaming, "I want to be in charge. I don't want Spirit to be in charge." Small ego doesn't understand how profoundly it is loved. The mantra, "Let go, let God" isn't heard from this space. Small ego says, "Listen here, knucklehead, I'm *God*! Who exactly do you think you are?" I'm not even sure that small ego knows Love.

The idea that small ego doesn't know Love is new to me and it is an extremely important question. I feel there is more than just nonjudgment as a connecting link between these two seemingly separate worlds and I'm not sure I know what that is right now. I feel I'm in very new territory.

Small ego is the cognitive understanding of the seven year old child, totally steeped in fear, trying to function in a huge world it doesn't understand. In some ways I feel the cognitive and emotional functioning level of small ego is even more like the seven month old trying to deal with separation anxiety, thinking, "I can't physically see the adult (the caretaker). The caretaker has disappeared." The unprepared baby is left to fend for itself and the baby is scared. Its survival is at stake. Even the seven year old child level of understanding doesn't have the cognitive skill of abstract thinking.

Many adults still function at this child level of understanding. Many adults, for a complex multitude of reasons, never even get to the abstract thinking level. This is one of the many reasons I believe a college education is so imperative.

Research emphatically shows how much the student's analytical reasoning ability changes from freshman to senior year. A person with less reasoning capacity and less emotional maturity hasn't yet questioned the assumptions learned as a child. This person doesn't yet think of themselves as: worthy, confident, capable, loved unconditionally, a creation of Infinite Love.

I'm not saying any of this as a criticism of anyone. I'm saying this as a profound reminder of gratitude for the life I have. I'm saying this as a profound reminder of the need for compassion! I'm detailing these realities because they have to do with Empowerment. When we feel empowered, confident, prepared, we can choose to live from Nonresistance. We understand we live in a highly ordered, beneficent, Spirit Identified Universe that is by Its Nature Unconditional Love. This difference in cognitive functioning level hugely impacts our understanding of ourselves, our relationship with the world we share with others, our definition of *God* and prayer, and I believe, our capacity to function from Nonresistance, AKA Trust!!!!! As I may have already said that is just one reason I am so devoted to education, spiritually and cognitively. We cannot function at our highest level when we're steeped in fear. We are certainly not Spirit Identified when we're steeped in fear. I believe our very nature is Spirit Identified. The key is we have to be aware of It before we can live from It. That's where I want to live from, and I want to help others live from It if they choose to do so!

Still later I hear these additional thoughts. Ego development is a necessary and important normal psychological process. The baby is learning it is a separate entity from its parents. It is not always an easy process. Parents didn't always get the nurturance they needed to be healthy. They don't necessarily have the skills to pass on the idea that we live in a beneficent Universe. They may not have experienced that reality.

What I call Spirit Identified Ego is when we have spiritually matured beyond our childhood woundedness and our limited understanding of who we are and how the world functions. It takes Self-Confidence, Awareness, Compassion to live from that expanded Nonresistance! That is our true identity! I believe that with all my heart!!!!

Study Guide Questions for Chapter 4

1) What does it feel like to consider being nonresistant to all of Life? Do you have a sense of what qualities are necessary in you to come from Nonresistance? Will you go there?

2) What does it feel like to love yourself and all others unconditionally? What fears do you have to face in order to live from that level of Love, Confidence and Awareness?

3) Do you have a sense of the healing your body~mind~Spirit experiences living from Nonresistance? What are you aware of as you consciously choose Nonresistance?

4) Will you imagine living from the affirmation: "**All** things are working together for my Good, and I am working with them in the Love and Wisdom and Power of Spirit."

5) What issues are coming up for healing as you consciously move through this process?

6) What does it feel like to consider that Nonresistance is Divine Love in expression? Will you make room in your Life to consider yourself as Divine Love in expression?

Chapter 5
Transition Point: Moving From Duality To Oneness Consciousness~

I added this chapter when I felt I had already completed this book. Today I realized I needed to say more about the process of change involved in moving from duality to oneness consciousness.

How comically fitting yet again that Chapter 5 is devoted to change. In numerology the number five is the symbol for chaos and change. Consciously or unconsciously, moving from duality to oneness consciousness can certainly feel like chaos. I love the Chinese understanding symbolized in their writing. The Chinese character for change is both crisis and opportunity.

Do you see how important our point of view is? As I was considering these ideas I saw the image of glaciers moving over glaciers! As beautiful as that image is to me I am aware of the white hot intensity that burns when colliding worldviews are moving against or across each other. It's uncomfortable!

Being awake to the change process feels more like we're gently flowing downhill in the current of a wonderful river, the Infinite Flow of Life. Being asleep to the change process feels more like the heaviness and heat of glaciers moving over us.

Here's why! These two systems of thought~feeling~being are completely different value systems! I'm sure you see that by now. The calling cards of duality are: fear; power outside of us; don't question authority; lack of self-confidence; guilt; shame; blame; avoiding responsibility for our thoughts, words, actions; and we can't know for ourselves how to lead~live our lives! Duality says we're at the whim of a capricious, unknowable Entity outside of us. We're left to wonder how to please~appease this wrathful Entity that has power over us!

Oneness says: We are Love Incarnate. We **are** this Entity in expression. We are Confident, Self-Aware, "response-able." We are this Power Source in expression. This Power Source within us, *as us* in expression, is completely knowable moment to moment in the Now! We are *God* Mind in expression! This Entity is Love Incarnate and so are we!!! We are to actualize this Oneness *in toto*!

I remember a clear, comical, chaotic and cosmic expression of this moving awareness process in my life. I was 26 years old. I was doing my graduate school internship in Psychology in the Contra Costa County Hospital near my graduate school. I was very aware that in one moment my first client was going to walk through that door and I was supposed to "do something." I saw immediately that my clients were quite willing to give me total authority and responsibility for their lives. My clients felt that I had all the answers; that they knew nothing. One client even insisted on calling me "Doctor" and I kept saying, "I'm a graduate student doing my internship. I am not a doctor." I did not want to claim power nor title that I did not currently have. I did want to help my clients empower themselves. I did want my clients to know I believed in them and I trusted them. I put the "response-ability" in their laps. I profoundly understood that my clients needed someone who would listen, who would care, who would help them empower themselves!

My knowingness has always been that we are capable. We are "response-able." I have no interest in having "power-over" someone. My interest has always been empowering myself and everyone to know that we have everything in us that we need in Life. That is why I left the church of my birth the moment I went away to college and was responsible for my own decisions. I knew myself!!! I realize what a profound Gift my Self-Awareness is! Believe me that is not lost on me!

I fully realize that others do not yet have that same Self-Awareness. Here's an example of how that plays out. I realize it terrifies some people when I tell them they have power to change their lives. My statement doesn't match their belief system. They feel overwhelmed and frightened by this new idea. They don't yet have a clue as to how they can change their lives for the Good. They still believe that power is outside of them. I have experienced this dynamic time and again in working with people. They somehow hear empowerment as a blame statement; as a criticism of what they are doing and have done. My intention is nothing of the sort but they sometimes don't hear my intention over the din of their fear.

Here's an example that blew my mind when it happened. Many years ago, with her permission, I took a person out of a chaotic situation into a peaceful, safe situation. Her response to my action was to attempt suicide. When I visited her in the hospital I asked her, "Why did you do that?" Her answer was, "I never felt so peaceful in all my life." I was shellshocked! Instead of seeing this situation as an opportunity to build a new life based on Peace, she must've seen the situation as a crisis, feeling she couldn't duplicate this process on her own! She couldn't or wouldn't see that help was all around her.

That kind of dynamic bothers me to this day. There's such a profound self-hatred at work when people choose to stay disempowered! They want to continue in the mold of being told what to do and then reacting to it or against it if they don't like what they hear. I don't understand that choice. It's painful to watch! It is in times like these, past and present, I must remember, "It's their story, not mine." I must remember the true meaning of a phrase I have come to truly appreciate: "One step forward, two steps back." I hear that now as the process of building a new pattern of unfoldment in my soul, rather than the self-defeating pattern it has been taken to be meaning we're getting nowhere. I don't believe it is possible to get nowhere. Our lives are always forward progressions toward full recognition and actualization of Fulfillment, our true Nature!!! How fitting yet again that I begin Chapter 6 with words I wrote many months ago that fit perfectly into how I ended Chapter 5 just now!

◆ ◆ ◆ ◆ ◆

Study Guide Questions for Chapter 5

1) Look at some situations you once called crisis and see how they turned out to be life-changing opportunities. What allowed you to name it opportunity?

2) What spiritual quality within you allows you to move from duality to Oneness? If it's difficult what is keeping you in duality? Can you allow Grace to love you?

3) What awarenesses are necessary in you to move forward without blame or judgment?

4) Do you call yourself an optimist or a pessimist? How does your label serve you or not?

5) When you see and feel pain within you and all around you how do you allow pain to serve you so it doesn't immobilize you? If you feel immobilized will you allow the spiritual gift of Imagination (Vision and Strength) to show you the way through?

6) How do you feel about moving into an exponentially greater Love for yourself and your world? What in you makes that move not only possible but exciting?

◆ ◆ ◆ ◆ ◆

Chapter 6
What Is Quantum Prayer?

Quantum Prayer is quite simply Empowerment! Quantum Prayer is a statement of our Source connectedness. Quantum Prayer is a state of being and a statement of being. Quantum Prayer is a state of being connected with the interior Life flow of our being, **as** our being. Empowerment means "power within." The Power is within us! Nowhere else! It's stunning when the clarity comes forth fully embodied. This awareness now feels like a slambang clarity. It couldn't be clearer to me now that prayer is **totally** an inside job; a simple alignment with the sanctity of our being. We simply want to know we have some kind of effect. We want to know who we are. We want to feel powerful. We want to be a true reflection of, and expression of, our true Nature, Spirit Personified. Looking within we empower ourselves each moment. We realize there is nothing in the outer that is going to change our lives. That's **our** job. We realize **we** are eternally empowered because the Power Source resides within us! **We are** Its expression! Our job is to use this awareness. Keep it quickened!!!!

We have to get over the idea that we can give over our authority and responsibility to something else or someone else. Our life is our job! We are not children! We are empowered entities in our own right! Flowing out of this awareness it is incumbent upon us to take back all the projections we've laid out on the world saying, "It's your fault." Any projection we cling to is our acknowledgment that we are still blind to how we are creating our world.

When I am willing to take back all projections I'm saying, "I am fully empowered, fully "response-able" to answer Life's call, Life's questions, in the present moment, with full Clarity, Love, Wisdom and Power! I am not a victim!!! I am *God* Love now and always!"

I wrote a poem in the 80s, "The Lightness Of The Dark Side", expressing my understanding that "the dark side is simply the aspects of *God* we do not yet know." Do you see how this process flows directly into my definition of Confidence? When we fully actualize moment to moment that we are *God* Consciousness Itself how can we be anything but Confidence? Not just confident, but Confidence Itself!!! I really love this state of beingness because it speaks so clearly to me of the "still quiet voice" of I AM, our true Nature!

When I don't understand what is happening or why or how I created whatever is happening in my life it is imperative not to take it personally! It is imperative to see the moment from a detached, nonjudgmental, almost comical, perspective saying, "Well, isn't that fascinating!" It is imperative to understand that the projection process comes out of the belief in duality, where we bought the idea that somebody is doing something to us! We think we have to psychologically defend ourselves against blame so we blame someone else. We have missed the point that there is no enemy! There is no need for defendedness when there is no fear!

Quantum Prayer is my awareness, "I am response-able! I can learn from this process. I see *God* Presence in this process. I give thanks for this process. I realign with a greater clarity. I create a different reality." Quantum Prayer is my reminder, "It's already answered! Relax. Calm down. I am with you always **as you**!" Quantum Prayer is my fully embodied awareness that I am an individualized expression of *God* Consciousness Itself.

This fully embodied awareness is a state of beingness; not just some activity I do when I want something. It is so important that we all know this! Quantum Prayer is a state of beingness that recognizes we are something! This awareness that we **are** something involves consciously living from a profound state of Gratitude, Joy, and Awareness. Quantum Prayer is a discipline of body, mind, soul and Spirit. By discipline I mean a study, not a chore! Anything done as a chore is a split from oneness. Anything done as a study is a joy. Anything we do in a state of Joy is truly empowered prayer. Embodied Joy recognizes we are all we need! We have all we need within us because we are expressions of It, Inexhaustible Source.

One example of Quantum Prayer that I've recently been reintroduced to on successive nights in two different settings is HeartMath. I love HeartMath. HeartMath is another one of the new sciences showing us we have all we need within us to be all we were created to be. HeartMath is a process of realigning with the natural coherence we experience when our heart and mind **is** aligned as one. Our heart actually has a brain of its own. Our heart brings all other body systems into coherence when our heart is leading the way. By focusing on appreciation we can change even the most explosive kinds of emotional situations.

The research being done on how our bioenergetic fields affect others is staggering. Our prayers even change the biosphere fifty miles above our head! This research is showing us we really can bring Peace to our world. This is what I call 21st century prayer! It doesn't even take a moment in time to recognize how profoundly different this definition of prayer is from the Western begging, pleading notion of trepidatiously approaching a recalcitrant, judgmental bearded white man in the sky image of *God* that most of us learned.

I believe that every thought, word, feeling, action, belief is a prayer. This is important when we consider that today's science as cited in the movie, "What The Bleep Do We Know?", says we think about 50,000 thoughts a day, and most of them are negative!!! Thankfully science is also showing us that positive thoughts are hundreds of times more powerful than negative thoughts as noted by Rev. Dr. Michael Beckwith in a 2008 interview in Unity Magazine. I take this information on the power of our positive thoughts as yet another expression of how incomparably generous the Universe is. This Life Energy that many call *God* is by Its very Nature expressively joyful, expressively continuous. That is the very nature of Energy; to expansively, expressively give.

That is why it is so important to realize that Energy gains momentum as It expresses. That's why what we do, what we say comes back multiplied. What we give out comes back multiplied!!! That's why it is so important to be conscious. Our thoughts, actions, beliefs, feelings, words represent our relationship with ourselves and the world we inhabit.

I love Rev. Dr. Michael Beckwith's distinction between what he calls thought and mentation. He says that mentation is simply our unconscious regurgitated pattern rather than a conscious prayerful mindfulness. Mentation is an example of what so many call prayer. So many people call their hand wringing worrying prayer! That is not prayer. That activity is simply an unconscious, disempowered, disembodied habit pattern that will bring more worry, more negativity, more disempowerment!

Prayerful mindfulness is a completely different fully empowered, embodied process. Prayerful mindfulness reminds me of the kind of attention the yogi saints and mystics give to life. Marianne Williamson said, "Don't expect to get the same kind of results in listening to a 15 minute CD that the yogi saints get from 8 to 9 hours a day in prayer."

That reminds me of another one of my favorite stories from Reader's Digest years ago about the boss who spent his first four hours of each day in meditation. His workers asked him, "How can you afford to do that?" His answer was, "I can't afford not to." The real kicker in that story is the boss got more done in his four hours than his workers did in their eight hour day. This story so clearly shows the power of concentration that an active prayer life gives us.

The flip side to that story from the 70s is the current Emotional Freedom Technique process seems to be a very real shortcut to meaningfully changing our thought patterns as we erase the brain pathways of the old destructive thought~feeling patterns we learned as children. I have witnessed some of these transformations myself and they are dramatic. People with lifelong medical problems are healed on the spot when the underlying issue is tapped out, that is, accessed and released. It's very powerful to witness this process. I've experienced EFT myself. I practice EFT every day. I consider EFT to be a form of Quantum Prayer. EFT recognizes we have direct access to the Divine, and that we **are** the Divine in Expression. EFT refreshes me in body, mind and Spirit. I experience Joy and Peace, direct expressions of *God* Consciousness.

I titled this book "Quantum Prayer: It's Already Answered!" because I truly believe what quantum physics and all the new sciences are now showing us. We are being shown the physical reality of what the mystics have been teaching us for years! The Cosmic Void, another name I use for *God*, is not empty. The Cosmic Void is the All In All made visible. This space holds everything within it. Anything that ever comes into existence is already present in the Cosmic Void. The Void is Energy. The Void is Intelligence. The Void is everything that ever was or will be. Our awareness of our oneness with all this Presence and all this activity allows us to call into being what is already there!

In the EFT process we use the set up phrase, "the same Stuff that created the Universe created me" as our recognition of oneness with *God*. I use the wording, "The same Love that created the Universe created me." This wording personalizes my relationship with the Universe and helps me feel my oneness more. The vastness of this Void expresses the nature of *God* to me. I call this Cosmic Void Infinite Love. We are Infinite Love made visible. When we tap into this Identity the power of our prayer exponentially increases. We are in tune with our very nature. We are in alignment with the endless power of the Universe within us, **as** us! We are unlimited! We have embodied our limitlessness! That is quantum prayer!!!

Before I leave this section on prayer I also want to address the concept of "singing is praying twice." I don't know who said that originally but I absolutely feel and know the reality of that activity. I have experienced the power of singing while using EFT. Using this combination exponentially amps up the power of my prayer work. It brings a level of Joy that I do not feel while just saying the words. Singing something as simple as Alleluia moves me into a powerfully peaceful, joyful state of being.

Singing activates a different part of the brain closer to our limbic system, our emotional brain. This Truth further highlights the Abraham teachings of the power of emotion in our lives. Singing is a feeling-based process. For me singing is an example of Quantum Prayer because singing so powerfully acknowledges our oneness with the Divine **as** the Divine. Singing words helps me feel the reality of being Spirit made visible. When we feel oneness we are fully activated, fully engaged **as** Life in expression!

———◆ ◆ ◆ ◆ ◆———

Study Guide Questions for Chapter 6

1) What was your childhood definition of, understanding of, and relationship with *God*? How did it serve you or not? What are your current answers to these questions?

2) What is your definition of prayer now? How does it serve you or not?

3) What is your relationship with these words: responsibility and "response-ability"?

4) How do you relate to these ideas expressed as Quantum Prayer?

5) How do you feel about identifying yourself as Confidence Itself?

6) What do you understand about this statement: "I want _____ but I don't believe I can have it? Do you understand the outcome of that split in Consciousness?

Chapter 7
Co-Creators! Not Passive Recipients!

The co-creative concept is so foreign to much of Western spirituality and yet it is the very essence of the embodied spirituality that means so very much to me. Co-creation is the very essence of my concept of Quantum Prayer. As spiritual beings, one with the Divine Creative Principle that is the very Life of all creation, it is essential to understand what this concept means to an empowered prayer life.

Co-creation is a very active process. Co-creation means we are in tune with the Creative Principle **as** us! Co-creation means we understand that by the power of our thoughts, words, feelings, beliefs and actions we are creating our life and our world. Co-creation requires deep listening, impassioned faith, inspired following of Guidance received, and Confidence in our identity and the purpose for our existence. Co-creation is qualitatively different than the "begging, pleading, we're not worthy, we're totally separate from" creation process. Co-creation is not a hand-me-down world. Co-creation is a Joy-filled, active, "response-able", creatively aligned participation with the very Essence of our spiritual human beingness! Co-creation means recognizing we are Intelligence. We are Power. We are Creativity Itself.

It is exponentially important to distinguish the difference between receptivity and passivity! Passivity and receptivity are qualitatively different states of being! Passivity is a completely disempowered, disconnected state of being. Receptivity is a vitally alive, listening, participatory, activated state of being. Receptivity is a fully empowered, connected state of being, recognizing ourselves as expressions of the One Source of all creation, our *God* Self.

Passivity is living from the "*God* can't do for us what we're unwilling to do for ourselves" state of being. Another example of passivity that drives me wild is when I hear people say, "Whatever *God* wants for me" as though they're petrified to make a choice in their own behalf. I want to ask, "What exactly do you think *God* is and where do you think *God* is?" I want to ask, "What happened to your sense of Self?" I'm not trying to be sarcastic or insulting here. This is another place where I experience how beaten down people are and it's agonizing.

Receptivity is a completely different state of being than passivity! Receptivity is a state of beingness that **knows** *God* speaks to us, through us, as us! Receptivity is the activated state of Flowing Inspiration. Receptivity is living from the state of being expressed as, "Do that which brings you Joy and I will take care of the rest." Receptivity is opening out into the exponential vastness of all that the Universe is, knowing that we are seen, heard, and valued! Receptivity is an impassioned Faith that opens out into the willingness to hear messages from Spirit like, "I want you to go to the edge of the cliff and jump off and I will catch you with feathers." I experienced both of these messages in the last three years when I was asking the Spirit within me, "What is mine to do?" I wanted a sign. I wanted to actively participate and to listen to the Guidance I knew I would receive. The answers were so profound and so immediate I cried. I was overcome with Joy and Love.

The co-creative receptivity process is another place to address our empowered relationship with fear. The Universe is so vast it makes perfect sense to be afraid. There's so much to do. There's so much to know. There's so much to be! It doesn't mean let fear stop us from doing what is ours to do and be. Letting fear stop us is a death sentence to the intrinsic Joy of our being! I have a favorite phrase from my Mystery School training at Diana's Grove, a retreat center in southern Missouri. The Mystery School at Diana's Grove is a year round program specializing in personal empowerment, leadership development, and spiritual community building. The motto one of my teammates came up with in our Rites of Passage year in 2001 was, “Do it afraid.” I love that motto! It means do all that is ours to do with faith, passion, and excitement.

Astrologically speaking, I’m a double Gemini. Gemini is an Air sign. Air signs are associated with communication. Air is expressed in many of the 12 houses in my astrological chart. That means there is an intense focus on expression and communication in my chart, and therefore, in my life.

The comedy of this is that I am also very shy and private in many ways, and yet I am totally devoted to public service. I’m sharing this to say that these fascinating personal dichotomies speak so profoundly to why I love the motto, “Do it afraid.”

I am the quintessential Gemini. I most definitely have two sides and yet they are very well integrated into my public and private life. I feel this is one reason I so passionately understand the difference between passivity and receptivity. I am a quiet private person in many ways. I am an introvert. I gather information through the internal processes of intuition and perception. I enjoy being alone. I am not a passive person. I follow the Guidance I am given. I know I am here to do and be all that is mine to do and be. I live by a personal motto I’ve lived for years, “I don’t want to die while I’m still alive.” By that I mean I do not, I will not, sit back passively while a deadening fear snuffs my life out! I am a leader. I’m passionately devoted to service. I am passionately devoted to being a change agent. I am passionately devoted to being Love In Action. Yes, I feel fear in these roles. I also feel the excitement of making a difference!!!

Another way I feel this emphasis on Air in my astrological chart is expressed in my life is my devotion to quietness and prayer.

Physically and astrologically the element of Air is related to breathing, communication, and prayer. Breathing, prayer, and communication are receptivity processes.

I find it exceedingly fascinating to consider the possibility that when our breathing is constricted we do not hear well! I have repeatedly experienced that people do not physically hear when they are emotionally afraid! Effective breathing is a quiet process. Effective prayer is a quiet process. Effective hearing is a quiet process. Perhaps a better word than quiet is relaxed! We must be receptive to ebb and flow. We must be free of constrictions! We must be conscious! We must be embodied!!!

Prayer and breathing are also co-creative processes. There is an equal exchange that happens in order for both of these processes to be optimally effective. When we are constricted in fear our breathing is affected. Our hearing is affected. Our prayer life is most definitely affected! That's why it is so important to be aware of our breathing when we're praying or meditating. When we're afraid we stop breathing for longer periods of time than we should. That's not healthy physically or spiritually. When we're afraid we instinctively revert to patterns of hypervigilance or extreme passivity. This is by no means receptivity. We cannot be effectively co-creative when we're afraid. We're co-creative alright! We're creating more fear. I doubt that's what we want. My point is be spiritually awake! Be conscious. Be receptive. Then we are effectively co-creative!

In the spirit of synchronicity and the comically amazing way Spirit works, this very moment I just learned another fabulously relevant example of the impact of receptivity. I just received a "Miracle Minute" telephone message from Rev. Mary Manin Morrissey as part of her Life Mastery course. Guess what her topic was? The power of yawning. The article she read to us addressed how the power of yawning impacts our whole system.

The spectacular sentence she related was that research finds that the chronically tired and the psychotic hardly ever yawn! How about that as an expression of the power of receptivity! Yawning increases the flow of blood to the brain, expands our diaphragm and loosens our jaw. Our body is saying, "Please wake up! Please be embodied! Please take in the Life Force you need to be connected." This isn't a blame statement. Just another fabulous example of receptivity to the Life Force in the expression of the Air we need to live. The more embodied we are the more connected we are.

◆ ◆ ◆ ◆ ◆

Study Guide Questions for Chapter 7

1) How do you feel about your body? What do you suppose that has to do with your method of prayer?

2) What is your relationship with passivity and receptivity in your own life?

3) When you get really quiet how do you feel? What do you do with quietness?

4) How do you feel about identifying yourself as Creativity Itself?

5) What so excites you that you're willing to "Do it afraid!"

6) What is your Life mantra?

◆ ◆ ◆ ◆ ◆

Chapter 8
Why Do We Ask?

Asking is our "I'm ready. I accept" signal. I remember clearly when this awareness came to me three years ago. I was overjoyed. I was asking some huge questions about what I am to do with the rest of my life. The answer I received was clear as a bell and immediate: "Do that which brings you Joy and I will take care of the rest!" I cried in recognition of the Love and Joy I felt. Asking is our statement of awareness of our oneness with *God* as an individualized expression of *God* Essence.

Asking is our awareness that our Good is already and always present and that we are indeed worthy of all blessings! Asking is our ready signal that says, "I'm ready to be the Joy of the Pure Being I AM!" We ask as a statement of Faith in the unending benevolence of the Universe. Just as the Genie in "The Secret", says to us, "Your wish is My command." Asking is our statement of readiness for the demonstration of the blessings that are always awaiting our invitation! Asking is our statement of our recognition of ourselves as Co-Creators in the process. Asking is our statement that we are present to the process. Asking is our affirmation of who we are!

I cannot possibly overemphasize the difference between asking from a state of oneness and asking from a state of duality. I clearly know the humiliation of asking from a state of duality. Its calling card is unworthiness: "You're not good enough. You couldn't do it yourself." All of these states of being are from fear, from the dualistic, judgmental "*God* outside of us" image that so many of us have grown up with in Western culture.

I think it was about 30 years ago I had an experience that signalled for me that I was moving to a different understanding of myself and my *God*. I had previously seen the experience of being brought to my knees as humiliation. I realized that what *God* saw instead was me being on my knees in reverence.

This realization is akin to the "trick" question, "How many people did Jesus heal?" The answer is none. Jesus so identified with his *God* Consciousness he saw only wellness rather than illness or any need. My realization of this All Abiding Love was a turning point where I realized I was releasing the judgments that I had introjected, taken in, and claimed as my identity. I was changing my definition of myself and in return I was changing my definition of *God* and my relationship with the *God* of my being.

I have continued on that road these many years later and I feel like I'm at the edge of a hugely expanded horizon of what I am and what *God* is in me, and as me.

It's a thrilling time. I'm not the only one making this move and it's so very joyful to experience this change in me and all around me. It's powerful. I'm grateful. Especially in the last three years I have had **so** many conscious experiences of asking from a place of oneness, and **isness**, and the answer has been so spectacularly immediate and huge I was either laughing or crying in joy and amazement and gratitude. One of these examples from this past three years is when I was asking, "What am I to do in my life now?" Everywhere I went during that time I kept hearing about Jesus' miracle of the loaves and fishes. I know that synchronicity is the Universe's signal to us that something spectacular is about to happen. Suddenly I understood how Jesus' miracle of the loaves and fishes happened. I was sitting at my computer doing my writing when I had the experience of being Jesus! I saw myself holding the loaves. The loaves immediately filled out as soon as they were broken. I understood that in giving we receive. I understood "My Faith is sufficient unto the day!" At that very moment the mail carrier walked up on my porch and deposited something in my mailbox. Later when I checked my mailbox I found a job offer for a position I didn't even know I was eligible for! I was so astonished I cried. What I also immediately understood is even though the money was spectacular that was not the job I was being guided to take. I was being guided to take the job that came through as, "Do that which brings you Joy and I will take care of the rest." It was a job paying far, far less and still I knew it was the right decision. Consciously I had no idea how this was going to work but I **knew** it would work!!!

It was in this process that I once again experienced the power and process of tithing. I was working so much in order to pay my bills I was getting sick from the pressure. I knew I needed to change something. I heard Spirit say very clearly, "You have to work less in order to do more." I understood this to mean I needed to start working one day *less* a week. I didn't know how that was going to work. I just knew it would. To compound this Faith paradox even more I began "tithing plus" (tithing an additional 25% of my tithe). I heard this inner dialogue where one voice said, "Are you out of your mind? What are you doing?" The other voice said, "Go ahead, kiddo. I got you covered. I understand exactly what you're doing!" Tithing at a time I was making 40% less than my previous job and yet my expenses were the same, was, and is, a huge act of Faith and yet it works! I knew the Power of my act of Faith. By my decision to "tithe plus" I was saying, "Thank You in advance for my Understanding. Thank You in advance for Your Presence I AM in me, **as** me!!! I am willing to live **from** Oneness." My feeling of pressure **immediately** left. Giving from Faith, Joy and Thanksgiving activates the Universe's incomparably gracious and spectacular responsiveness. The point of saying all this is I'm repeatedly learning how incomparably I am loved. When I ask in Faith I **always** receive! Often I receive so far beyond what I've asked for my only response is tears! The tears are a heart-connected response to the Love I've received! I've since had countless demonstrations of this asking~receiving process. I keep following Spirit's Guidance and answers **always** come and I follow the Guidance received.

Another learning in this asking process is: Spirit's only way to be visible in a physical Universe is *through us*. So when I need something from the Universe I know it is going to come through direct actions of my own, one of which is asking a human for help. This might seem like common sense but I truly feel it's something bigger. Being willing to ask involves both Self-Confidence and Trust. Asking means profoundly understanding how much we are loved. In a world based on fear and being small we are so conditioned to hide and to be defensive. We are loved **so** much greater than we've ever imagined. It's high time to open to the everpresent blessings. As the Genie in "The Secret" says, "Your wish is My command." In order to receive Its benefits we must plug into the receiving mechanism. We must open to receptivity, knowing we are undoubtedly worthy. We must understand our part in activating the receiving process. We must ask!!!

◆ ◆ ◆ ◆ ◆

Study Guide Questions for Chapter 8

1) What are you ready for? What are you willing to ask for and accept in full Radiance?

2) The Genie says, "Your wish is My command!" Now's the time! What's your wish?

3) Spirit manifests through you! What are you co-creating?

4) What is your relationship with tithing? How willing are you to be prospered?

5) I have repeatedly found that people are so willing to be generous. What is your relationship with asking? What does it tell you about your relationship with receiving?

6) Giving and receiving are inextricably linked. What is this saying to you?

7) What do you feel about my loaves and fishes story? How is it relevant to your life?

8) What is your definition of miracle? What does it tell you about your relationship with Oneness or duality consciousness?

Chapter 9
What Do We Receive?

When we ask in a state of readiness, knowing the answer is already present and fulfilled in Spirit, the number one blessing we receive is a greater understanding of who and what we are and what our relationship with the Infinite is. We receive the entirety of the Universe every time! That is the nature of the givingness of the Universe! I realize these are bold statements. I believe them nonetheless. I certainly don't always understand that I have received the entirety of the Universe every time. That says something about **my** readiness, **not** the Universe's givingness!

My Self-Love, Faith and Understanding are still growing. I believe we always receive the information that tells us what more we need to do and how to keep on going if we will just have the heart to stay tuned into the Infinite Presence always within us, all around us, as us!

By "have the heart" I mean two things. I mean be spiritually ready for the Universe's infinite magnanimous overflow of blessings. I also mean be aware that our heart is the physical organ that aligns all the other pendulum systems in the body because our heart is the largest pendulum. The simple explanation of this is that when the heart is congruent, that is, at peace, it brings all the other systems of the body to the state of congruence, or peace. That means all systems are on the same page! All systems are working together for our very highest good! There is peace, not stress. The body and mind are on the same page! The brain and the heart reflect the same peaceful, congruent rhythm. Our brain waves, respiration rate, heart rate, digestion, and elimination are all in peaceful rhythm.

This information from HeartMath is very important. It is the physical reality of the spiritual reality which means: Be open!

Be receptive! Without being open we **cannot** receive! We must learn to love ourselves no matter what we **think** is happening in the outer. We must maintain the inner awareness that the outer is simply a reflection of the inner whether we're conscious of that reality or not. We must stay connected within!

Here's an example of the importance of staying connected that absolutely blew my mind when I first heard it. In the HeartMath experiments the research showed the impact of five minutes of focused hatred. The immune system was depressed for six hours!!! Five minutes of focused love boosted the immune system for six hours!!! Can the tradeoff be any clearer? What impact do we want to make on our body~mind~Spirit continuum? This kind of research has huge implications for the field of psychoneuroimmunology where the body~mind~Spirit continuum impact is so clearly seen. Attitudes are powerful!!!

Another absolutely fascinating example of giving~receiving body~mind connection is seen in the research of Dr. Masaru Emoto's work with water crystals. Our bodies are 70% water. He said he feels that Alzheimer's is the result of sadness and loneliness. These feelings somehow collect aluminum in the system and the result is Alzheimer's. What I understand this to mean is when our overriding feelings in life are sadness and loneliness somehow there is a major heart disconnect. Our major spiritual lesson is how to love!!!

From separation consciousness we might be saying, "Are you kidding?" No, I'm not kidding! I've been to separation consciousness too. I'm learning to forgive myself and others when I unwittingly find myself and others in this place of separation. I remind myself daily it's part of the "Clarity journey." It's part of clearing out all the resistance, all the "less than" messages, all the fear-based messages that do not serve me or you! In HeartMath language, we bring ourselves to a congruent state by focusing on anything we appreciate. When we feel all tensed up or out of sorts we focus on appreciation. We breathe through our heart. Our breathing changes. We begin to relax. Our systems all come into balance. In a matter of moments we bring ourselves to congruence, to peace.

Every moment we spend releasing resistance leaves more room for Clarity, Peace and Joy! There is no greater work to do than leaving room for more Goodness to come into our lives! I believe with all my heart we are here to receive the unbounded beneficence of the Universe. The Universe recognizes Itself in us! That's the way I understand the verse, "It is *God's* good pleasure to give us the kingdom." Through our connection with our Divinity in us, **as us**, and all around us we have a fully embodied Understanding that Heaven is here now! Heaven is simply a state of consciousness, a place of Peace and Bliss here and now each moment! We're the only ones who separate the connection by our fear and our belief in our unworthiness. We don't have to do that ever! We can be brand new each moment!

◆ ◆ ◆ ◆ ◆

It's many months later since I wrote this chapter. It's rather fascinating to come back to this thinking I was going to add a big understanding and to see it already written in the first lines. The greater understanding of who I am that I **have** received in this writing process is it couldn't be thrillingly clearer to me what it means: "If it is to be it's up to me." I absolutely **know** that what I hold in heart and mind can't help but come into being. It's already present in Spirit. It is absolutely in the visibility process. I truly truly know that Truth. That's quantum reality! Quantum reality is the expression of the exponential beneficence of the Universe. If anything is taking longer than I thought it just means: "Be still! Have Faith! Keep going! Your effort is powerful and undeniable! I AM listening." I hear this as a Love message to myself from my Self. I will not shortcircuit my efforts in any way. That would not serve I AM. I am here to serve and express I AM!!!!!!!!

What I will amend in this section and add to it is that along side of making room for more Good by releasing what no longer serves us we are to build the parallel Universe of what we do want. It's already present in Spirit!!! By our attention to It we bring It forth into visibility. That effort goes hand in hand with the necessary release work. When we ask we receive Guidance always. If we don't understand the Guidance I believe it means we're either going too fast or we're not ready. Sometimes it seems comically clear what we are to do and our answer is, "You've got to be kidding!" I believe that answer means we don't understand and accept our Greatness! Time to look again! We **are** Greatness!!!!

◆ ◆ ◆ ◆ ◆

Study Guide Questions for Chapter 9

1) What is your giving~receiving ratio (meaning can you give and receive equally well)? If your answer is no do you understand the impact on your Life?

2) What have you not yet forgiven? How is it serving you to not forgive yourself?

3) If your life were vastly more peaceful than it is now what would you have to change about yourself? What do you hear me asking here?

4) What is your Life purpose? How will you fulfill It?

5) What is your definition of Heaven? What does Heaven feel like to you?

6) How do you grow through separation consciousness? How does awareness serve you?

7) Are you ready and willing to accept that the Universe loves you unconditionally? What old teachings do you have to give up in order to allow more room for already present Love?

8) What event did you identify as a crisis and now recognize as a gigantic blessing?

9) How undefended (open) are you willing to be? What blessings do you imagine?

Chapter 10
Prayers Are Always Answered!

Prayers are never unanswered! Yes, another bold statement! To say that prayers aren't answered is to say that *God* isn't present! I personally don't believe that ever happens. It's different to say my prayers weren't answered in the way I expected, needed, or demanded, than to say *God* isn't present. In prayerfully addressing this huge question and mystery one of the questions that means the very most to me is, "How far can we bring the Magnanimity of Spirit into the physical realm?" This question fits into many chapters of this book. I especially want to address it here because this is where our understanding of Magnanimity and Presence is most needed.

Dr. Jean Houston did a great service for our human community with her work on what she calls the four levels of reality. As I've said, "There's often so much more going on than what we see in the physical realm." I have worked with the four levels of reality principles since 1997. I first heard of these principles at Diana's Grove. These principles have helped me become much more conscious of what I am doing and feeling from moment to moment. In this work on the four levels of reality Dr. Jean Houston speaks of the physical, emotional, mythic, and essence levels of reality. I call the Essence level the Spirit realm. One way I work with these principles is to ask myself, "What archetype, original pattern, or Principle am I invoking right now?" "Am I invoking fear as the Essence of my reality or am I invoking Love, Generosity, Prosperity, Life?" I remind myself I have unlimited choices. This is another one of those places where I feel it is **so** important to understand the physical and Spirit realms are simply two expressions of the same place. I do not separate myself by making a value judgment that one is better than the other!!! "Different than" is much different than "better than" or "lesser than." Oneness is so qualitatively different than separation! This is our human lesson of a lifetime.

Sometimes we are profoundly deaf and blind to the blessings that abound because we limited **how** our prayer was to be answered! I'm including myself in the acknowledgment of missing the blessings that abound. Our point of view is the key! Are we seeing *God* Presence or Absence? When we're unconscious we don't see how our emotional reality is impacting us. We don't see how we've built things out of proportion for perhaps our whole lives. We've built such a mythic level of illusion and separation we don't or won't see or experience anything else but separation and one disappointment after another. We live from such an unconscious level of self-hatred anything that doesn't fit our definition of unworthiness can't come into expression in our world. Intentionally or unintentionally so much of our childhood conditioning teaches us to be afraid, to live in fear. We cannot experience Joy and fear in the same moment. If nearly the entirety of our identity is fear there's no room for Joy and Peace. We wonder why our prayers haven't been answered. We don't even realize that we haven't left room for anything else! We don't realize that every prayer **is** answered. We don't realize that every thought, word, feeling, belief, action is a prayer! We don't realize that by steadfastly filling our lives with negativity we are prodigiously creating even more negativity! Prayer is Energy. Energy is always present! Therefore, answers are always present! When we become as adept at framing the positive as we are at framing the negative we will create a much different reality in our lives. Being willing to frame the positive is an Awake process. That is why I frame Self-Confidence as a spiritual Gift. Self-Confidence is **not** just an emotional process. Self-Confidence is the awareness that we are the Divine in Expression!!!

How we are feeling about our lives tells us how ready we are to see answers everywhere! Answers can't come into being until we're receptive. They're already here but we won't see them until we're ready! The "response-ability" is ours! We *are* that powerful! Yes, we are! We make a colossal mistake by thinking that our prayers have to be answered in the way we ask them to be! We forget the perfection of the process in the larger picture when we try to orchestrate how things are to happen. We are "playing *God*" when we do that, rather than recognizing the *God* Presence as us! It doesn't mean our loved ones won't die. It doesn't mean we won't experience pain. There's a difference of understanding between time and eternity. It's the same reality, just a different expression. Our need for immediate gratification, immediate answers, getting exactly what we wanted when we wanted is an expression of our fear and our lack of understanding. All this is an expression of our belief in *God* Absence, and our fascinating human propensity to need to tell *God* what to do! It's disempowerment! It's fear. We think some outer Entity is creating our world or answering our prayers. We don't realize **our** co-creative activity.

Especially for those of us raised in the Western hemisphere where there is such a disembodied unawareness of so much it may not be a surprise that when asked what they wanted in life 40% of the people asked couldn't even articulate what they wanted!!! I find that both sad and astonishing. In order to spiritually progress we must holistically understand there's far more going on than what we see in the physical realm.

Reality has many levels: physical, emotional, spiritual, soul, mental, mythic, essence, maybe even more than that could be named. When we've prayed for healing and someone dies, perhaps it looks to us as if our prayers weren't heard and our prayers weren't answered. I do believe death can be the healing. Death truly can be the release into Joy! We have no way of knowing what understanding a person may have experienced at the moment of their death. We have no way of knowing what another person's soul journey is. We may not even know whether the person believed they could be healed or not. If they didn't believe they could be healed then they can't be.

I go back to the age old statements about answered prayer that I quoted earlier, *"God* can't do for us what we're unwilling to do for ourselves," and *"God* helps those who help themselves." I hear those statements saying to us, "When we are in alignment with our highest good our prayers are answered in the way we ask them to be or even better. When our prayers are answered in a different way we must know that there's more at work than just the physical realm.

This understanding helps us especially when we are dealing with the grief of losing a loved one in death. Few of us can see on the other side. Those who can see the other side report that the person is blissfully happy. We may be sad but they are not! In all of this it is important to remember *God* is Love, and we are co-creators with *God*, as *God*. We are never alone. Our prayers are powerful. Our prayers are Energy Itself. All of these factors are in play each moment in the physical and the nonphysical realm.

Another piece of this picture is to realize that in Spirit it is always Now. Our Good is always with us whether it's visible or not. We're not used to living in that awareness. We think there has to be a time delay. When we think there has to be a time delay that's exactly what we create! Yes, we are that powerful! That's our dualistic, time-based thinking at work. As said in "The Secret" it's often a good thing there is a time delay. It's the Universe's gift to us saying, "Are you sure that's what you want to activate?" The way we so often hear a time delay is, "You don't like me. I'm not good enough." When there's a time delay there's also too much room for fear.

Another side of the same separation or delay consciousness is that when we're in alignment with who we are, and we're not used to being in alignment, the answer is so immediate we're stunned. We may even reject the good, feeling that can't possibly be for me. That happened too quickly. On the other hand, when we are used to being in alignment our response is gratitude. That's a fun place to be. That's an important place to be! I'm getting used to it more and more.

All that I've said here about answered prayer is one more reason why the Abraham teachings are so important to me. These teachings have added so much depth and Joy to what I already know and they are presented in such a comically nonjudgmental way it's easy to sit back and see myself and just playfully say, "Yes, thank you, I see myself better now!"

The key piece of the Abraham teachings for me right now is the lessons on being aware of what vibrational level we're at, knowing that if we're in tune with the level of Joy we're creating room for more Joy in our lives. If we're at the level of despair we're creating more despair. At that point Abraham says to us, "You have to move from that place. You have to get some relief. You have to move to a different vibration in order to change your experience."

A powerful way any of us can bring some relief is to simply say, "I love you" to ourselves or someone else. Every statement is a prayer. Every statement is a recognition of our current definition of *God*. Saying "I love you" is our way of saying I'm ready to be open again instead of closed down and in on myself. Saying "I love you" is our way of saying I'm ready to move from blame to awareness. Moving from blame to awareness dramatically changes the tone of our prayer life and our experience!

That's one example of why being aware of our vibrational level is so important and why it is so important to consciously remember that we are co-creators, not passive recipients. Passive recipients simply create more of the same unconscious life they're living. There's no Joy there. We must be willing to ask ourselves, "What prayer am I asking to be answered? At what level do I want my prayer to be answered?" In order to say I want Joy I must recognize I am Joy! If I don't feel I am Joy It cannot come to me.

When I'm in a place of "less than Joy" I recognize there are ingrained unconscious messages of self-hatred, unforgiveness that I'm focused on so I must make room again for something different than that message. It is perhaps the most empowering lesson to understand how our painful experiences can serve us. We can choose compassion for self and others, or we can choose embitteredness! Need I say the experience is qualitatively different? I'm not trying to be funny here. The choice is Life or Death to our conscious beingness.

I feel we're constantly being asked by the Universe, "How much Bliss are you willing to welcome into your life? I have more than an abundance to joyfully share with you, my Beloved." Being willing to live at this level is being willing to move into the Avatar level, our fully actualized Spirit Identity, the complete embodiment of our true spiritual nature! I believe with all my heart that that is what this physical life is for, for each and every one of us. So, I ask us all, "At what level are we individually and collectively willing for our prayers to be answered?" So mote it be! So may it be! So it is! Amen!

◆ ◆ ◆ ◆ ◆

"Every moment is simultaneously a prayer, a blessing, and an answer." The answer is **always** in the question! That's why it is **so** important to pay attention to how we frame the question! I'll end this book by playfully and passionately asking: "What shall it be?" My answer is: Quantum Prayer: It's Already Answered! Quantum Reality: It Already Exists! Thank you, precious Universe! I love You so very much! So mote it be! So it is! Namaste! Alleluia!

Study Guide Questions for Chapter 10

1) What most fervent prayer is as yet unanswered for you? Nonjudgmentally, what are you learning in the process? Is there some place where fear is in the way? How will you be changed when your prayer is answered?

2) What prayer is answered beyond your wildest dream? How are you different as a result of this answer to your prayer?

3) Who answered your prayer? Do you understand what I'm asking?

4) Describe the *God* of your being and your relationship with this Entity?

5) How do you feel about an approach~avoidance relationship with Life? From Nonresistance how do you see this approach~avoidance relationship serving you?

6) How do you understand my comment: "Every prayer is answered."

7) When you say to yourself, "I am Joy", what is your feeling?

8) How do you feel about the realization: "There's always more!"

Chapter 11
Living As The All In All or Living As Judgment?

What Shall It Be? I thought I was done saying everything that needed to be said for this project. The awarenesses keep coming that say, "Not yet, kiddo!" So here's what I'm hearing that needs a chapter of its own, not just a weaving in, out and through ten other chapters.

I'm quite sure by now it is clear to you that some of my very favorite spiritual concepts are: Faith ..... Generosity ..... Magnanimity ..... Confidence ..... Empowerment ..... Flexibility ..... Exponential Love ..... Transformative Joy ..... Oneness!!! Underlying these Principles is my knowing: the Universe is unfailingly benevolent!!! We are loved!!!!!! We are precious!!!

Equally clear to me is not everyone shares this knowing. As a result of that different knowing, Life is defined and experienced differently. The consequences are enormous. One glaring difference is how humility is defined and experienced. Humility as Oneness is a calm confidence. A simple thank you when complimented is humility as Oneness in action. This definition, experience, and understanding of humility is very important to me. I don't see the true essence of humility being understood or honored in our Western society. This is relevant to the entire content of this book because it impacts our definition of ourselves, *God*, and prayer! Humility as Oneness is: Self-Awareness! Self-Acceptance!

As a bridging the gap comment perhaps this is the place to raise a question that means a great deal to me. In another one of Unity's most famous affirmations, "It is *not* I but the Christ within that does the work" I see the difference between Unity's panentheistic understanding of *God* and my pantheistic understanding of *God*. My wording is, "It **is** I *as* the Christ within that does the work. This is an important difference. It speaks to the very heart of my entire message. My understanding of humility is to accept with a calm confidence who I am. I am *God* in expression. We all are. That is our fully actualized Identity. In Humility, from Humility, we accept the full "response-ability" to awaken!!!

In Unity's 19th century wording I hear duality. I hear "I" being labeled as **only** separation ego rather than my definition of "I" as *God* Identity Ego. I hear duality again in the devaluing of sense-based consciousness. I view the senses as sacred. I view the body as sacred. The senses and the body are both expressions of the Divine in the physical. I believe we see, feel, and hear the Awe and Wonder of *God*, as *God* through our senses. For me, the body is simply Spirit made visible. I don't hear that meaning in Unity's 19th century wording.

I want to be very clear yet again that I greatly value the teachings of Unity. I also value my understanding. The very purpose of prayer is to empower people to know their true identity. I want to bring us into what I consider the full understanding of who we are in the 21st century: Body~Mind~Spirit Oneness, *God* Expressed. It's not enough to acknowledge that *God* is *in* us as Unity does. That idea doesn't reflect the Majesty of who we are. We are Spirit Essence. Spirit cannot create anything less than Itself. What we do with our life is up to us. It is our highest holiest calling to live **from** our full all out Awakeness, our fully actualized *God* Self. That is a Joy-filled journey!!!

It's just days later since I wrote that last paragraph. I've had yet another "here's where the rubber meets the road again" level of understanding. The Oneness definition of Humility is quite simply Self-Confidence. We fully recognize and accept our *God* Self Identity and we consciously choose to live from It. We know who we are. We know and fulfill our divine purpose. We recognize and revel in the Truth that Spirit cannot and does not create anything less than Itself. That's not heresy. It's quantum reality!!!

I fully realize that is not the prevailing understanding but I sure intend it to be!!! It **is** certainly the Knowing I choose to live from. To do anything else is to deny my *God* Self Expression and I **won't** do that!!! I know that as I live **from**, **and as**, the Light I AM, I change the world. All who have eyes to see and ears to hear receive the message. It's not about controlling anybody. It's about living **from** the Truth of my being. If and when anyone else comes along is up to them. I release the timing of the outcome. The timing of the outcome *for others* is not up to me!!! If there's a contrast between my timing and their timing I understand that *God* functions as contrast so we can see ourselves in action. I'll do my very best to view that difference from a comical and cosmic perspective... seeing the Whole picture!!!

In the 80s I wrote a poem, "Freedom: Our Ultimate Responsibility." Freedom is a spiritual conscious discipline. Freedom is **not** the unconscious process of license!!! Freedom means recognizing and cherishing the Truth that we have an impact on others. I cannot live someone else's life. That is not for me to do. I can be a Light bearer!!! I **am** a Light bearer. I so "get it" that the level of consciousness we rise up into is up to us. Now I fully understand the words I heard earlier this year, "Rise up into healing." I was hearing *God* Self speak to me. I intend to listen with all my heart, mind, body and soul.

When we need a 2x4 contrast here's our reminder. From separation duality the imbalances of "power over" and "power under" are glaringly obvious. Humility is expressed as a need to refuse a compliment and put ourselves down. A belief in our inherent unworthiness *permeates* this expression of humility. This isn't humility!!! It's an enculturated unworthiness that can, and **so** very often does, go to the extreme of self-hatred. That is why I can't say often enough how completely that definition and that process misses the point of the Majesty we are!!! That is why Empowerment, *God* Consciousness, is **so** important to me!!! Our Nature is Exponential Love and Light!!!

From duality we are taught from day one: It isn't polite to think well of ourselves. Some attitudes and actions that follow this thought pattern are passed off as admirable spiritual practices: austerity, self-flagellation, poverty. These practices share a common belief: We are inherently evil. We are sinners from day one. We deserve to be punished!

The body is even defined as evil. Spirit and body are defined as qualitatively different! In this process the mind and body are cut off from each other. Arrogance flows out of this disconnection. Fear and ignorance underlie this disembodied, disconnected process. It is important to recognize that arrogance and ignorance are simply flip sides of the same coin. It's the same value system. It is important to understand there is a meanspiritedness about arrogance. Arrogance is based in fear! Arrogance is an artificial buffer for a control we know we do not possess because we're rooted in fear.

There is such a spiritual poverty in all of this, poverty can't possibly be an admirable quality. Jesus said, "I came that you might have Life and have It abundantly." He also said, "Greater things than these shall ye do." Neither one of these statements came out of a belief in separation duality. Jesus didn't teach separation duality. He taught Oneness.

My point in all of this is a profound knowing that Life is supposed to be exponentially Joyful! We cannot be Joyful when we're judging, when we're disconnected, when we don't understand who we are! That's why I'm introducing, highlighting and expanding some of these principles here.

It is **so** very clear to me: Faith is the ground of my being!!! It is **so** very clear to me: Confidence comes out of being rooted in Faith!!! I am absolutely positive: Unconditional Love is the Essence of the Universe. I am absolutely positive: We are **intentional** expressions of this Unconditional Love.

———— ◆ ◆ ◆ ◆ ◆————

It's a little bizarre to me that 11 chapters into a book there are still so many huge ideas to introduce and expand. Today it occurred to me: Uninterrupted Energy is a rhythmic wave. Interrupted Energy flows like a kink in a hose: either completely stopped or jagged. This interrupted energy pattern can also be a spiritual and electrical expression of resistance. Rather than make an enemy of resistance like we often do when we feel caught in duality I liken resistance to an opportunity to strengthen and grow. I see it as an opportunity, not a crisis. It's a powerful experience to see pictures or EEGs of these principles. One picture shows us Oneness. One picture shows us judgment. They're powerful reminders of our everpresent Choice points all along the way!

Here's where the rubber meets the road again. I love the story of an angel being present with us at birth. This story so graciously and meaningfully helps me be at peace with the pain I've experienced in life. The story says: "Before we were born into physical being we made soul agreements with the Universe to learn certain lessons. Then just before we were born an angel touches our lips and we forget the lessons we agreed to learn."

Personally I think it would be overwhelming if we saw our whole lives laid out before us (assuming we can even do that or that that is how life unfolds). Personally I think our lives are primarily blank slates that we color in as we go! We change the final outcome of our lives by the decisions we make in each moment. Do you understand that? Potentially, we live millions and millions of lifetimes in one life experience, changed and colored in by each decision we make each moment. Remember multiple possibility worlds?

This all goes back to my beginning comments about my definition of *God*, prayer, and us! It is so clear to me now that Confidence brings about our conscious awareness of our Oneness. Fear brings about separation! No doubt about it for me! My question for me personally continues to be: "How willing am I to permanently release the old messages and live the new Life that is already, and all ready, in Spirit?"

This is the cliff I'm ready to jump off into now! This is the cliff Spirit invited me to jump off when I heard Spirit so clearly say to me, "I want you to go to the edge of the cliff and jump off and I will catch you with feathers." The new message I heard just now as I typed this is my inner Spirit saying, "I'm giving you the feathers as our symbol that you're ready to take flight." My goodness, I can feel the Love!!!!! I have chills sitting here typing this!

◆ ◆ ◆ ◆ ◆

Contrarily, I immediately feel the embodied separation in the pit of my stomach when I'm judging and blaming, not understanding how any moment fits into the perfect whole of my life unfolding. Metaphysically, fascinatingly enough, and I know it's no accident, the stomach represents our Power center. Negative judgment is a physical and emotional experience. Negative judgment so clearly tells me I don't trust myself. I don't love myself! I don't trust *God*. I don't see the whole picture. I'm lost in fear. I have just cut myself off from my Power! I have just cut myself off from *God* in me, as me, through me!

When we judge and blame and get lost even momentarily in the past or the future it is because we have given more attention to judgment and blame than to the Now moment, not loving what is, as is! Don't just take my word for it! Try it out yourself!

Loving what is, as is, is the hallmark of Trust (code name Faith!). I feel it takes enormous Faith and awareness to love what is. We must constantly be aware of the larger picture; the whole of our beingness, our *God* Presence as us! We must Love ourselves, Trust ourselves, embody Faith, as is!!! We must be present to ourselves, to unfoldment! I don't think it matters which one of these archetypes of spiritual awakeness we identify. They all mean Presence: Trust! Generosity! Magnanimity! Flexibility! Resilience! Love! Faith! Confidence! It's all Presence!!! **Any** identity is enfolded in the All In All.

——— ◆ ◆ ◆ ◆ ◆ ———

It is very important to make one distinction. Negative judgment is not the same as the spiritual quality of discernment. Discernment is Wisdom. Wisdom means we are heart and mind connected as Spirit Incarnate. There is no negative judgment from that place. There is only a fully embodied Love~Wisdom.

Another key concept in this definition process is the word surrender. There's a world of difference in meaning and intention in Oneness perspective and duality perspective. Surrender in duality means, "You lose." Surrender in duality has "power over" tones: "I beat you. I won. I'm better than you."

Surrender in Oneness perspective means open out into the widest, highest, holiest view. Surrender means trust the Universe's exponential love for you! That's how I hear the words so popular in Unity teachings: "Let go! Let *God*!" Let go (of our conditioned focus on fear, self-hatred, doubt, anxiety, pain. You're safe. There's no need to be guarded and afraid). Let *God* (Unconditional Love, Faith, Strength, Vitality) do Its perfect work in us, through us, **as us**! **Know** that we are all enfolded in Love, by Love, ***as*** Love every moment! Separation thoughts don't exist when our focus is on our *God* Oneness, Isness!!!

An interesting twist on this popular Unity phrase is: "Let *God*, and let go!!!" I love that thought. Start with Love first. Leave fear out of the picture. Recognize our *God* Identity!!!

———◆ ◆ ◆ ◆ ◆———

The awareness that came through crystal clear this morning is: Life is not so much a journey to the mountaintop as it is a letting go of all the judgments that keep us from knowing we're already at the mountaintop. We **are** the mountaintop!!! Enjoy the ride. Enjoy the Presence!!!!!! That's why I love glaciers so much!!!!!!! A few days ago I was wrestling with this idea of Oneness, paradox, and duality. I knew I wasn't quite getting it all. Now I see clearly. The glacier holds the entire picture for me. The majesty and beauty of who we truly are **and** the immovable object when we don't allow Love to melt our frozen emotions. We already are Spirit made visible. Revel in that Consciousness!!!

———◆ ◆ ◆ ◆ ◆———

As I further consider these ideas and the spiritual awareness process involved in living them out I come again to my understanding that if I'm only seeking perfection to avoid pain I've missed the point entirely. I don't need to pursue Perfection. It's already and always present!!! That's the very basis of my understanding of quantum prayer and quantum reality: Resting in the Presence!!!! The underlying lesson I continue to learn is that Wisdom comes through learning from the mistakes I make. We live in a physical Universe. The Abraham teachings remind us that we learn what we want through desire. Desire arises from a need or a want. The Law of Polarity reminds us that in a physical world there are always opposites. My point here yet again is don't make lack an enemy. Embrace it and allow it to be the simple impetus we need to move toward what we want, knowing it's already present in Spirit and we **are** capable of bringing it into the visible realm. Again it's a matter of our focus. Focus on what we want rather than reacting to what we think we don't have! Love ourselves in the process. Cultivate Nonjudgment and Nonresistance! Leave room for Joy, Humor, and Appreciation. Walk gently through life.

Another word that is so important to me to define from Oneness is meek! In Western psychology and spirituality meekness is seen as weakness. It is nothing of the sort to me. Weakness doesn't fit the meaning I feel in Jesus' words, "The meek shall inherit the Earth." For me meekness embodies the calm confidence needed to accomplish every goal with Ease, Peace, Joy and Enthusiasm, embodying Heaven on Earth.

I wanted to know the Aramaic translation of this passage. When I saw the words, "The meek shall inherit the Earth" in the Aramaic translation at first I was surprised. Then I realized the passage truly meant what I felt it did because Jesus does not speak of us as weak. He speaks of us as having the kingdom (the Consciousness) of Oneness within us!!!

It is important to note here that the Bible was first written in Aramaic (Jesus' native language), then translated to Greek, Latin, German and English, in that order. It would be nothing short of fascinating to have each one of these translations side by side to see how they compare. The point is some languages don't even have a word for the word another language is using so when translation happens meaning can be dramatically changed.

This is another one of those places where I hear so clearly St. Francis of Assisi's words: "There is nothing so strong as gentleness and nothing so gentle as real Strength." In Western spirituality and psychology we have been **so** ingrained with fear-based defendedness we don't even know or see the real Truth of Life. As sad as that makes me I do realize it is yet another polarity that leads us home to ourselves through our mistakes. So yet again I'm reminding myself and all others with eyes to see and ears to hear: Walk gently through life. It is a precious unforgettable journey. We are here to be a blessing and to receive a blessing. That is the Law of Love: our true Nature!!!!!!!

Another word to understand from Oneness is "wait." Wait means to serve!!! Service is a very active, awakened, conscious process. Wait is **not** the passive connotation we have learned in Western spirituality and psychology. Yet again this is so very important for an empowered understanding of who we are. This passive connotation of wait feeds into the mistaken connotation of *God's* time. We must reawaken to the realization of knowing that *God's* time is **always** NOW!!! The only time anything happens is in the present moment. Remember Jesus' statement: "the fields are already white with harvest."

Here's another word that is the very definition of quantum prayer and quantum reality: Limitless!!!!! By our very nature we are limitless. Our lives qualitatively change when we embody the Truth of this prayerful reality. When we realize the power of our words we don't say things like: "I live on a fixed income." The **only** way for our income to be limited is if we feel it is; if we think it is. By our very feeling and thinking then we have made it so! I highly doubt that's what we want.

We're back to square one: Our feelings and thoughts are powerful. Be conscious of what we're feeling, thinking, saying, doing. We are Spirit in expression. We are limitless. That **is** the Truth. **Anyone** can live that Truth into being in the physical world!!!!! That's what we're here for each moment. We have to be willing to take the blinders off. We have to be willing to think, feel, and live for ourselves. We have to be willing to be empowered: "Response-able." It is our Life. We must be willing to live Life to the fullest; to step into and live into the fulfillment of the promise and the potency of who we are.

You might be asking, "Well, isn't that just natural?" My answer is a "both~and" answer! Yes, it **is** natural and yet people are so buried in fear they are afraid to even remotely step into this new Life. They choose to stay asleep thinking that's safer! It's not! It's simply a living death! The key is: Live our own lives. Don't take theirs on; it's theirs.

Nonjudgmentally do what is ours to do and allow them the space to do what is theirs to do in their own time. It is not our responsibility to "fix" them. That's a death trap. We are to shine our own Light. That's empowering and helpful for us all!!!

I cannot leave this section on the oneness~duality paradigm without specifically highlighting the implications of one of the most famous Bible verses: "Where two or more are gathered in my name, there am I in the midst of them." This passage has always been interpreted to mean where two or more people are gathered. I don't believe that's the intended meaning of this passage. That would imply that praying by ourselves is ineffective and that *God* isn't present when we're praying by ourselves. I don't believe that at all!!! The meaning I see is speaking to our state of beingness. When our sentient heart~mind is united as one we are in alignment with Presence, **as** our *God* Self Incarnate. We are not divided against ourselves. We are in alignment with the entirety of our beingness. That's the meaning I see in this passage.

In a similar vein two other ideas express the "*God*-outside-of-us paradigm" that is **so** disempowering and so disheartening. "Of ourselves we can do nothing." My answer to that is, "Of course we can do it. It's our life. Who else is going to do it? It **is** our "response-ability. We are more than capable of handling the situation. We are God Identified Ego!!! We are more than equal to the task at hand."

What I have never heard dealt with in the "Of ourselves we can do nothing" statement is the underlying identification with fear as our **only** identity. When we self-identify with fear we put ourselves outside of the solution!!! We are divided against ourselves. We have cut ourselves off from the solution. The solution **is** still there but we have put ourselves on the outside!!! We have in effect said, "I want this but I don't believe I can have it." We have forgotten our Love~Wisdom Self Identity in us, as us, **is** the solution!!!

The other statement is similar. "I am powerless over this situation." These are disease model statements. These statements put the solution outside of us. Systems of treatment have been built on this idea. These statements are dualistic rather than holistic. It's time to recognize **we are** the solution!!!

In this move from duality to Oneness Consciousness it is so very clear to me we must **consciously** reevaluate *every* word we've ever learned. We cannot build a conscious Life on unconscious assumptions. We must be consciously aware of what cornerstones we're making the foundation of our precious Life. Life is a time for Awakeness. Passivity is **not** Consciousness. Passivity doesn't serve us. We must evaluate and re-evaluate the paradigms we've unconsciously set up for ourselves. We must now and always ask ourselves instead, "Will it be Unconditional Love or immobilized fear? Will it be the Love~Faith continuum or will it be the fear~self-hatred continuum?" The contrast couldn't be wider!!! The impact couldn't be more significant.

The point I'm making yet again: Words are powerful. Concepts are powerful. Intentions are powerful. What is our intention in what we are saying and doing? It is a precious, precious matter of consequence that we live consciously!!!!!!! A playful, sincere caveat here. Even decades into this consciousness raising journey it is astonishing to realize and experience the intensity of the judgments that come up in me as I continue my effort to live from Nonjudgment. The Law of Polarity is here with a bang. When we say we want to release judgments they flare up brighter and bigger than ever as if to say, "I'm going out with a real bang if you're serious about releasing me."

Have I ever been living through this process lately. I'm not trying to be funny here. I'm dead bang serious. It's like the frozen crystallizations are scraping against every self-doubt I could ever have to let me know what moving into the next level of Self-Love and Clarity really embodies. I hear it as the next level of Trust asking me, "Are you really ready to respond from Love rather than anger?" It's a real Clarity journey to respond from Love rather than reacting from still unrecognized anger.

I keep experiencing this Truth. It's not comfortable. The point is: Recognize the meaning of the discomfort and keep going and growing. It does get easier. Trust, Confidence, Compassion and Appreciation grow in the process. Gentleness and Strength grow too. I know that for sure! As I've said before I keep learning what the Jesus level of Love involves. Divine Love is a hugely magnanimous spiritual quality!!!! Divine Love is Nonjudgment. Nonjudgment is a crystal clear Confidence, Clarity and Presence!!!!!!

That's the fascinatingly huge paradox. Love has no opposite!!!!!!!! Love is *God*. *God* has no opposite. *God* is simply Presence, Isness. We know clearly we're not in *God* Consciousness when all the opposites show up!!!! I never knew there could be so many opposites!!!!! Now I am being funny and real too!!!!! Time to breathe, relax, reset, begin again!!!!!! Enjoy the process. Relax in the Presence!!!!!! Life is precious. Enjoy It!!!!!!

———◆ ◆ ◆ ◆ ◆———

A few days later here's what awareness leaped out at me. Spirit has no opposite. It is simply Presence. It is simply Energy. The opposites we see and experience in physical reality are for experience, perspective, and discernment; not judgment! Embodying this reality is: Quantum Prayer, Quantum Reality, the fully actualized Avatar Experience.

———◆ ◆ ◆ ◆ ◆———

"Every act of judgment is an act of Self-hatred (hatred of our *God* Self)!!!" That stopped me cold! I've had that awareness before but this was more embodied than ever! I asked myself, "Is **that** how you want to live your Life?" I heard myself say, "I think not!!! Let me listen more carefully to what I'm doing."

It is very important to clarify here yet again that negative self-judgment is **not** the same as righteous judgment. Righteous Judgment is Wisdom. Wisdom is simply the clarity that comes from living from Unconditional Love, our *God* Self Identity.

Negative self-judgment is based on fear, our separation identity. Negative self-judgment is a reactive defendedness based on self-hatred and fear. It's another Abraham moment: Wake up! Let in Love! Be the Love I already am! Be Peace! Help others heal! If I can't or won't do it for myself do it for others. That very act of Faith and Generosity frees me! A related precious reminder: When I need something pray for others! If I keep negative judgment going I keep pain going. I keep self-hatred going in myself **and** others.

Every decision is a binary process: A "Yes" or a "No"! Sometimes our decisions are premature. We say it's an open and shut case. That's judgment. That's not seeing. It's not vision! In judgment breathing stops! In judgment Joy stops. Joy is breathing! Joy is an ongoing opening and closing: a continuous acceptance and release. Joy is Spirit connectedness! So~~ Our Choice: the All In All or judgment? It **is** up to us! If it is to be it's up to me! A precious matter of consequence!!!

This precious matter of consequence keeps getting clearer and clearer in the most surprisingly phenomenal ways. I continue to receive a more and more deeply embodied understanding of Nonjudgment. Nonjudgment is quite simply Unconditional Love. Nonjudgment is Self-Confidence. Nonjudgment is Nonresistance. Nonjudgment is the deeply embodied understanding of our Eternality, our Majesty, our *God* Self Identity!!! We have virtually no sense of this Majesty when we're hunched down in fear.

I experienced the hauntingly primal nature of my newly embodied understanding of Nonjudgment in a recent dream. Dreams have a way of bringing a clarity that's even more striking than words. In my dream a big gray cat and a tiny gray kitten were laying together quietly in the same box. All of a sudden the big gray cat clawed off the kitten's face and yet I could still see the kitten's face. I could feel and hear the kitten screaming and crying. I was holding its disembodied head. I was softly blowing in the kitten's face to comfort it. The kitten stopped crying. I woke up with a start!!! I didn't understand the dream. I heard myself asking the big gray cat, "Why did you do that? You're both the same." I wasn't angry. I was mystified. I wanted to comfort the kitten. Later I understood this is the primal nature of fear when a new reality is being born. One part doesn't recognize the other part. The big cat doesn't realize that the kitten is a product of the big cat's own life. I felt the dream was telling me, "Be gentle with the new Love and Wisdom that is being born in you." I further understood that fear tears us to shreds. Survival mode is primal. Survival mode is not logical. The heart and mind are torn asunder in fear mode. We must survive in the physical for further spiritual development to happen in this lifetime. It's not always a smooth ride. There is a primal drive for survival because there is a primal drive for the upward evolutionary spiral of spiritual development that is our very Life. Being motivated by fear brings one result. Being motivated by Love brings a different result.

I asked a friend what was her understanding of my dream. She said, "You have an understanding that comfort is beyond words. Comfort comes from the breath of Life."

◆ ◆ ◆ ◆ ◆

Study Guide Questions for Chapter 11

1) What is your feeling about the words: humility, meek, and surrender? How have these concepts influenced who you are and how you live your life?

2) How do you feel about the reality that you bring Heaven consciousness to Earth?

3) What is your harshest self-judgment? Who will you be when you change your attitude?

4) How do you feel about the idea that the Bible has been grossly mistranslated or translated to suit the prevailing political opinion of the day?

5) When you love what is "as is" how does your life change?

6) What is your understanding of, “The meek shall inherit the Earth.”

7) What is your feeling about the angel story in this chapter?

8) Who will you be when you are ready to live a new Life?

Chapter 12
Full Circle!

As we near the end of this shared project it occurs to me I must speak to the difference between linear perspective and circle perspectives. In an odd sort of way linear reminds me of Ouruborus' tale (and tail). In linear time there is the illusion that the beginning isn't connected to the ending (even though we're eating our tail). We aren't aware that the end and the means or the cause and effect are connected. It's different in the circle perspectives. The inherent understanding in the circle perspective is interconnectedness. There is no deliberate teaching that we are disconnected! There is a deliberate teaching of: immanence, respect, equality, community and cooperation. Interdependence is understood and honored. Connection is celebrated!!!!

In this context I would like to share two powerful experiences and understandings from my life. Even as a child I was a serious reader. I was constantly reading articles on: Medicine, Education, Psychology, Science, and Religion. I have always been a keen observer of people; always watching what they did and wondering why they did what they did. I particularly wondered about the extremes of human behavior. I was fascinated by psychopaths. I felt they weren't understood. I seemed to have some kind of unspoken understanding of the pain they were in that motivated them to do the horrible things they did. I think I must've been wondering, "How does that happen and what can I do to help?"

In 2001, during my Rights of Passage celebration at Diana's Grove leadership training program, a friend of mine said to me, "You have the widest definition of forgiveness I've ever heard." I was insulted at the time. Her comment was made during a very emotional, very meaningful time for me. Her tone of voice seemed sarcastic. It has taken me until 2008 to understand what I think she meant. I think she was saying, "Holy cow, I can't match your definition!" When I told a friend of mine that story today she said, "You see *God* in everything." I smiled in appreciation of her understanding of my intent.

It is absolutely my intention to live a fully embodied *God* Presence in each moment. This intention is the connecting link between my childhood awarenesses and my adult understandings. I am aware of the **huge** Faith, Confidence, Trust, Awareness it takes to stand in the middle of horrendous pain and stay centered. There is still so very much more I need and want to know about what it takes to look past the appearance; to see only *God* **as** every person, situation, and moment!

I believe this is the lesson of a lifetime to say the very least and I don't know that it can happen in just one lifetime. Magnanimity is such a **hugely** generous concept and beingness. I don't do it every moment and it's agonizing to me that I don't do it every moment. I so very much want to live totally from the Spirit of Generosity that we are created to be! I believe this willingness brings individual and collective world healing and transformation. I said to this friend today, "Forgiveness is the Grace when we can't come from Love the first time around."

I know it takes exponential courage to look a psychopath in the eye and love them and forgive them; to see their spirituality and their humanity! It takes courage to look anyone in the eye and say, "I love you. I forgive you!" I know people have been so deeply hurt by so many things in this lifetime. My lifelong intentions continue to be: Hold the Love, the Light! Transform the lives of all I touch! Educate, uplift, empower and transform the whole consciousness of the Universe. As I have said throughout this book I know that's why we're **all** here. I love the saying that echoes this Truth for me: "There's only two reasons to do anything: To be a blessing and receive a blessing." I know it is impossible to do one without doing the other. I know the power of Energy and Intention. That is why I define *God*, prayer, Life, you and me in the way I do. It is the very reason I wrote this book. Blessings of Joy and Peace to us all!

◆ ◆ ◆ ◆ ◆

Study Guide Questions for Chapter 12

1) What is your relationship with: Forgiveness, Grace, and Divine Love?

2) What is your understanding of where linear and circular perspectives intersect?

3) How has Faith served you?

4) How did you come to define yourself as an optimist or a pessimist? How does your definition serve you? What would change your definition of you?

5) What is your understanding of the concept: A psychopath is Presence.

6) How would Self-forgiveness serve you?

7) What soul agreements do you feel you made with this lifetime? How do they shape you?

◆ ◆ ◆ ◆ ◆

Chapter 13
Still More Is Coming Through

The Never-Ending Cycle!!! I keep thinking I'm done with this project. Such huge realizations keep coming through I know there's more to say. It's been weeks since I wrote those last lines in Chapter 12. I keep adding more in other places. These thoughts needed their own chapter. What came through me today is: the three most powerful prayers are embodied in three words: Thank you! Stillness! Presence!

All three of these words speak to me of the Presence of Joy. Joy is the state of being when all the noise stops; the noise of our disconnectedness. I am aware like never before of the power of Stillness! Stillness allows room for the already present Joy to flow in and through us each moment.

I am so very aware of what the absence of Joy feels like. I feel consumed by this awareness lately. It's very odd because I feel so very empowered. I am seeing in profound ways what judgment, resistance, disconnection feel like. I feel like I'm on the cusp of understanding like never before the healing power of Joy, Stillness, and Presence. I can see clearly how Jesus, the yogi saints, and all the avatars of all times do their work. It isn't work! It's Joy! It's Presence! It's Stillness! There is no work to be done! There's only beholding the Truth: Wholeness! Oneness! *God* Essence! Presence! That's what I know in principle! Now to be there........ to live there each moment...... and not regret when I'm not there. Just begin again brand new in each moment. That's the lesson!

Yet again I can see Ouruborus' tale and tail. The very point of this life other than living a life of Joy the very best we can is not to make pain the enemy! I don't like pain. It's uncomfortable and yet without it we would die at a **very** early age! There are true medical stories about people who are born with no pain sensors and they die very early because their bodies have no way to say, "I need help please!"

I realize right now pain isn't the problem! It's all the judgments we are **so** enculturated to heap upon ourselves so immediately. We're using our minds the wrong way. We must stop the noise! There's a difference between identifying something for descriptive purposes and communication purposes, and identifying **with** something! It occurred to me that the yogi saints, Jesus, all the avatars, all those living at the ascended level here on this Earth don't even need to identify anything. They have just one word: Oneness! They can walk through a pit of vipers and not have one molecule of fear. There's only Oneness! This is another place of Stillness. They're fully awake. They **are** Presence! So are we when we are awake! We are to appreciate each moment. We are to appreciate the blessing of any pain and make a new choice. No attachments! No noise! Just Peace! Stillness! Awareness! Joy!

This is the bliss of the circle perspective. Each moment is brand new. Around and around the wheel of Life~~ brand new each moment. It is a discipline. It is a practice. The purpose is living an embodied Joy! I do know that's possible. It is our individual and collective purpose. I further know that Joy is lived from a proactive Presence place. The pain that is separation is a reactive place. We're caught in old stuff. We reacted rather than grew through the lesson. We're still in the past, not in Presence! Knowing this as I do now the realization that comes through clearly is I want to cultivate much more room for Joy and Stillness. As confident as I am and as naturally optimistic as I am I know fully there is **so** much more room for Joy and Stillness.

I see Ouruborus' tale and tail in how we've been taught to vilify our emotions. In dualism, emotions mean we are vulnerable. Dualistic teachings say, "Be impervious to pain.

Be invulnerable. Be highly emotionally defended. Be very afraid. Be hypervigilant. It's the only way to be safe!" Research shows that emotions get stuck in the body when we deny our emotions. This is an example of the disconnect between mind and body that happens when we don't allow thinking and feeling to express as a unified whole.

Gestalt therapy is one of the many processes we can use to heal this unnatural division in who we are and how we function. Gestalt therapy helps us identify our feelings and connect with our feelings once again so we can talk with them and hear what they have to say to us and understand how they are here to help us. Gestalt therapy was the focus of my graduate school training in Psychology. These principles have been a huge gift in my life. I have had some powerful healing experiences when I have talked with the pain in my body and asked it, "What do I not yet understand?" Try it! You'll probably be surprised!

The picture I see right now is the straight line of our bodies as we stand. Our spine (symbolic of Courage) is the direct connection between our body and our mind (which we symbolically identify as in our head). I see Courage as one of the many names for our Spirit connectedness. Now I fully understand the words I heard 30+ years ago, "Beware of the pendulum as it swings from psychopathic rage to the "doormat mentality." There is a middle ground called Love." I didn't understand until now the picture I saw in my mind the first time I heard those words. How can Love be the middle ground between those two extremes? What did that mean?

Now I understand: The middle ground is Stillness! Presence! Divine Love! Not the wild emotional fluctuations back and forth while we're reacting to every outward change. The middle ground has many names all signifying Presence; knowing there is only Now; that centered detached place where Observer

Self bemusedly and amusedly lives, watches and responds. These names are all examples of embodied Presence. The holistic viewpoint honors the body as sacred. I honor the body as sacred. Our cultural worldview healing comes as we once again recognize the body is Spirit Individualized! Joy lives in our bodies! Joy is our Spirit connectedness. Joy is how Spirit expresses Itself in the physical.

——— ◆ ◆ ◆ ◆ ◆ ———

Dualism teaches us that we are to vilify the feelings in our body. We are to vilify the body~~ period! Dualism teaches that the body is identified with the Feminine which is viewed as negative, inferior to the mind. What I see clearly as I near the end of this writing project is the power of balance, and (pun intended) the balance of power! We are physical beings. We have a body. We have a mind. They are both sacred. They are both needed in a physical Universe. They are **both** expressions of Spirit.

In holistic relationship body and mind function as a unified whole. We are meant to experience Joy in our bodies. Joy is how we know our true Self. That is the awakening I see happening throughout our precious world. That is why I'm doing the work I'm doing. That is one of the many reasons I wrote this book. I am doing everything in my power to live from Oneness. I'm doing everything in my power to celebrate Joy, Stillness, Peace, Wisdom, Generosity, Flexibility; all the spiritual gifts of Oneness and Presence that I love so very much. I do believe as I originally said that *God* is everything. *God* is all there is. The paradoxes of the contrasts enfolded in that understanding will keep me endlessly, fascinatingly engaged until the end of my time on this precious Earth we all share together. We're on the road........ together!!!!!! Namaste!!!!!!!!!

◆ ◆ ◆ ◆ ◆

I find it quite cosmically fascinating that I am now writing Chapter 13 of this book. When I began this project I had no idea I'd be writing this many chapters. Do you know that 13 is the Goddess' number? That's why there are no 13th floors in any building. It is yet another way to vilify the Goddess, to hide Her influence, to deny Her very existence. Do you know the real symbolism of the number 13? Very simple. The Goddess traditions are Earth-based agricultural traditions based on the lunar calendar which has 13 months marking the Moon's 28 day journey by which farmers plant and harvest their crops.

As I said earlier in this book Goddess traditions are empowering healing traditions meant for everyone's benefit! Goddess traditions are community-based traditions. Goddess traditions have absolutely nothing to do with Satanism or Devil worship. Associating the Goddess traditions with Satanism is a deliberate teaching perpetuated by the patriarchal traditions to scare and control people. That's why education and empowerment are so **very** important to me!!! I want everyone to know they are powerful. I want everyone to know they are inherently good. As I wrote that last sentence I find it eerily haunting that we refer to Chapter 13 as a bankruptcy process. Goddess traditions are thought to be practiced by morally bankrupt people. I want to share a quite different understanding with you.

I am trained in the Reclaiming tradition of Goddess spirituality. "Reclaiming" refers specifically to reclaiming our relationship with Nature recognizing the Earth is a living system that we are caretakers of, not dominators! This is very very important. Dualism teaches us that the Earth is a dead mechanistic process; that our actions have no impact on the Earth. Nothing could be further from the Truth. Goddess traditions bring us back to our natural alignment with the Earth as a sacred living system. Goddess traditions are based on recognizing our oneness with Nature. We are creations of Nature! We are to honor our Nature, not suppress or repress It. Central to Goddess spirituality is honoring the body and sexuality as sacred! Embodied Joy is central to Goddess spirituality.

Empowerment is central to Goddess traditions. Oneness and equality are central to Goddess traditions. Goddess traditions are known and celebrated on every continent of this precious Earth we share together. What much of the world calls prayer, Goddess traditions call Magic. Magic is defined by Dion Fortune as: "the art of changing consciousness at will." Magic is central to Goddess traditions. Magic is a profoundly reverential conscious practice focused on a very specific intention. Magic is the process of linking with the Creative Principle of Life. Magic recognizes that Energy is powerful. Energy is creative! Energy is not something to be played with or treated lightly! Our thoughts, words, feelings and actions are powerful. As I have said earlier in this book Goddess traditions are not just for women or about women. Goddess traditions are for and about everyone and everything! Goddess traditions are based on cooperation, oneness, interconnectedness, respect for all life, and community building.

I am sharing this to bring healing, understanding and respect to our world. Finding the Goddess traditions was another experience just like finding Unity. In 1987 when I read Marion Zimmer Bradley's classic book, "The Mists of Avalon", the story of King Arthur told from the woman's point of view, I cried and cried and cried. At one point I felt like I had melted into the pages of the book. I felt like I had travelled directly to Avalon, right through the pages of the book. I felt I had come home to myself. Just like Unity I never knew such a tradition existed until I found it. When I found Diana's Grove in 1997 I cried every time I left for the first few years. I did a lot of crying because I was going there 10 times a year. I didn't want to go home. I felt so open, so safe, so welcomed, so connected, so whole at Diana's Grove. Thankfully I have long since integrated all those feelings fully into my life. I want that kind of openness and wholeness for everyone everywhere! Yet again that is why I'm doing and saying what I'm doing and saying!

———— ◆ ◆ ◆ ◆ ◆————

Yet again it is weeks later since I wrote that last paragraph. It is abundantly clear that I must address one more reason why I love the Goddess traditions so much. The Goddess traditions promote such a natural understanding and honoring of the fullness of the Life cycle: birth, growth, death and rebirth. We come from the Invisible and we return to the Invisible. For eternity we are Spirit. For a while we are Spirit made visible. A sweet thought just came to me. We have three sacred homes: Invisibility, Visibility, and the Present. Our Invisibility home is in Spirit as Spirit before we get here to this physical plane. Our Visibility home is our bodies. Our Present home is each moment. As we **fully** experience each moment we are free and clear as we are meant to be. We are fully conscious!!! No residue! No stuck pain! No regrets! This is quantum prayer and quantum reality. We are Power Itself: Present! Fully Awake! Joy-filled!!!!!! Ready for take off!!!!

This kind of Aliveness can't happen when we are sleepwalking through our enculturated death denying Western world. A death denying process is a fear-based process based on the idea: You lose. Dying is viewed as a defeat instead of a natural process. Releasing this fear is important. We cannot fully enjoy our lives when fear of death is running our lives. We cannot attain and maintain a beingness of quantum prayer and quantum reality when we're locked in fear. As Spirit Incarnate we are Power Itself. We are the Light of the world. We are the Love of the world. I know that with all my heart. Yet again that is why I am saying what I am saying and doing what I am doing.

———— ◆ ◆ ◆ ◆ ◆————

This is a very fitting place to highlight another difference between the holistic Goddess traditions and the dualistic patriarchal traditions most of us have grown up learning. I have had stunning realization after stunning realization about what "Getting my own way" means. Here's how some of these realizations play out in our physical and emotional reality. The patriarchal tradition is a competition-based win-lose system. I was thunderstruck a few days ago as I realized "Getting my own way" is a substitute for spiritual advancement.

I intentionally stopped for a few moments to let that awareness fully sink in so I could take in the full feeling of that awareness. "Getting my own way" is a surface level of living. It is saying, "Stop the pain. I need some relief. I'm at my limit of what I know." It's a fear-based, unconscious habit pattern of thinking, feeling, being. It is separation duality's way of being powerful. I am the only one present in "Getting my own way" mode. There is no thought of what anyone else is experiencing. No one else is even present in my experience.

This means I have no concept of the Whole! I am living from a "less than" reality, instead of I AM Spirit Identified quantum reality. There is friction and resistance in this "less than" place. The polar opposites living here are extreme agitation and severe depression. There is no Empowerment. There is no feeling of Grace, Ease, and Efficiency in this "less than" place. There is too much residue of the past and the not yet even actualized future.

As I was having these realizations I deeply felt, "Oh, this is what Spirit Identified Peace feels like. There is no residue. There is just great Grace, Ease, and Efficiency. No wasted motion. No wasted moment. Not a molecule of residue."

I realized like never before that in the "less than" place thoughts and feelings are disconnected from words. This is so very important. Communication and dialogue are not taught in the authoritarian patriarchal culture. I have witnessed time and again people who can't even speak their hearts and minds. They're so diminished by fear and unworthiness they can't even get the words out. Claiming their place in the world would mean being visible and they have learned that being visible is dangerous and frightening, so they're practically mute!!!

This dynamic comes out of the authoritarian culture that says any kind of a communication effort or any comment in our defense is heard as back talk. Feelings are not honored. Anger is especially threatening. Instead of being recognized as a natural expression of frustration where respectful dialogue would be appreciated and helpful, anger is seen as an affront to authority instead of as a wake up call to the authority figure to be responsible for their impact on others (all too often literal impact, as in physical, emotional, even sexual abuse). These dysfunctional communication patterns are all aspects of the "power over" dominator process of duality. As we continue our move further into "power with" and "power within" dynamics we heal these abuses and misunderstandings that come from "less than" perspectives.

From Oneness and Presence we are "response-able", fully capable of speaking calmly, confidently in the moment expressing our feelings, thoughts, wants, and needs. It doesn't mean we don't experience anger in the Oneness perspectives. It means we can speak calmly, respectfully from "I am" statements instead of speaking from blaming statements ("you are" statements). An example of an "I am" statement is: "I feel frightened when you scream at me." Instead of saying, "You sound like an idiot when you scream at me." We respectfully acknowledge our feelings and express our needs. Communication is the intention, not degradation, not shame, not blame.

In contrast, the polar opposites of Oneness are expressed as the deafening loudness of agitation when we feel we don't have what we need and the deafening silence of severe depression when we don't recognize who we are! In these agonizing "less than" places we are so pain-identified we often don't even know how to communicate. Living from Oneness brings us back to Wholeness where loving, effective communication is possible.

Quantum prayer and quantum reality bring us back to ourselves, our Spirit Identified Wholeness and Wellness, our natural state of being! From Oneness we know who we are and we know we can meet every need in perfect peace and order as we live from Spirit Identified Ego. We are at home in our world. We are comfortable with ourselves and our friends, known and unknown! We live, speak, and act from one of my favorite Goddess tradition mantras: "In perfect Love and perfect Trust." That is my intention in Life.

As I end this Spirit-filled, Joy-filled project I want to especially acknowledge the coming transformations of 2012 that signify the end of the 5,125 year cycle of belief in separation duality. 2012 signifies the full recognition of the Divine made visible in our physical reality.

In that Unitive Spirit I want to say again **why** I am so passionately devoted to the Thomas Gospel understanding of who we are! My understanding is that Spirit cannot and does not express anything less than Its Perfect Self. That is what I hear and so love in the Thomas Gospel. That is the understanding Thomas had of who we are.

In the dualistic Gospel of John what I hear is, "Here's the standard you are to emulate but you will never attain it." That is a double bind and any double bind is a card carrying recipe for craziness because it says you cannot access and express the Wholeness you are. As an educator, mental health professional, and simply as an empowered person that is **never** the message I want to give to anyone. That is why I am also **so** passionately devoted to healing this fractured sense of love and who we are that has been passed down to us as reality.

Right now I **so** fully realize that in saying, "Thank you, *God*", we are saying thank you to ourselves for taking honorable action in our own behalf. This recognition is truly the quantum leap that makes all I've said here so very real to me. I am **so** very grateful for this quantum leap in Consciousness. It is so very important to acknowledge that evolution is an inextricably linked physical~spiritual process. As we change physically we change spiritually. As we change spiritually we change physically. Each moment that interconnection unfolds in front of us and within us all. Now is our time to consciously leave so much more room for the expanded awareness of the Love, Joy, Peace, Faith, Confidence and Generosity we already and always are. We haven't even uncovered the tip of the iceberg of our exponential Greatness. There's nowhere to go but up!!! I'm so excited and grateful for the present and future unfolding in front of all of us!!! Thank you, Spirit, for Your everlasting unfolding Love expressed ***as*** us all!!! Namaste to us all!!! Thank you for sharing this journey with me, my dear readers!!!

———◆ ◆ ◆ ◆ ◆———

Study Guide Questions for Chapter 13

1) If Goddess traditions are new to you do you see these ideas and practices serving you?

2) Do the ideas so challenge you that they frighten you?

3) How do you feel about the idea that your body is sacred?

4) What would you consider doing to make your life a more intentional practice?

5) How do you feel about a worldview based on cooperation, respect, and community?

6) What is it like for you to consciously "stop the noise"?

7) Can you picture your life as you cultivate Stillness as a spiritual practice?

8) Can you picture your life as you consciously cultivate Joy, Flexibility and Generosity as your spiritual practice?

9) How comfortable are you with your whole range of feelings?

10) What do you consciously do to create a sense of shared community for yourself?

11) What is your relationship with Observer Self?

◆ ◆ ◆ ◆ ◆